Enamels

THE SMITHSONIAN ILLUSTRATED LIBRARY OF ANTIQUES

General Editor: Nancy Akre

Enamels

Susan Benjamin

COOPER-HEWITT MUSEUM

The Smithsonian Institution's National Museum of Design

ENDPAPERS

Enameling in process at the Parisian bronze foundry of Barbedienne, founded in 1839 at 30 boulevard Poisonnière. Both the *champlevé* and *cloisonné* techniques were employed; *cloisonné sur font* was cast, wire compartments included. This 1886 view of Barbedienne's *cloisonné* workshop, with finished vases displayed on a stool, shows unusually luxurious working conditions. Courtesy *Harper's* magazine

FRONTISPIECE

Four centuries of Chinese *cloisonné* enamels: a fifteenth-century shallow libation vessel with spout and ring-handle; a sixteenth-century Mei-p'ing vase, a shape seen in porcelain from centuries earlier but not in enamels until this period; a seventeenth-century box in the form of a duck; and an eighteenth-century Ting incense burner. Height of incense burner: 35 cm. (13¾ in.). Victoria and Albert Museum, London

For Laura and George

For their invaluable advice the author wishes to express her sincere thanks to Mr. Hugh Tait, Deputy Keeper of Mediaeval and Later Antiquities at the British Museum; Dr. Ada Polak, Deputy Curator in Britain of the Arts and Crafts Museums of Norway; Mrs. Ursula Robertshaw; Mr. Michael Gillingham of John Sparks Limited; Professor Gerald Benney; Mr. Peter McCabe; Mr. Louis Lawrence; Mr. Momin Latif; and Mr. Geoffrey Munn of Wartski Limited.

Printed in the United States of America

Library of Congress Catalog Card Number: 82-72761

Art Direction, Design: JOSEPH B. DEL VALLE

Picture Editor: DOROTHY SINHA

Contents

1 Introduction

Enameling has long been considered an outstanding example of human skill, for it is a refined and sophisticated technique. From earliest times it has been a craft patronized by the wealthy. Enamelers have worked for kings and emperors and their courtiers, for high prelates and the princes of the Church, and for rich and discerning collectors.

As manufacturing costs decreased with developments in technology and improved production methods, attractive enamels became available to a wider public in the form of tableware, trinkets and keepsakes; eventually enamel came to be used commercially and industrially—its hardness and its resistance to chemicals imparting a value in both kitchen and laboratory.

The advantages of enamel are many. The jewel-like brilliance of its vitreous surface is durable, and the colors in which it can be produced range from the bold and vivid to the subtle and pastel. It may be used to cover large areas, for example to produce a mural, or to form part of some tiny detail, such as an enamel "bead" in the filigree frame of a piece of jewelry. It can be fused to either precious or base metals.

The value of an enameled object is obviously affected by the cost of the metal on which it is based, but also depends on its age (early enamels are rare), condition and artistic merit. Many enamels are really pictures in miniature and the same critical standards apply to them as to a full-scale work. The artistic quality of an enamel is entirely dependent on the skill and creative ability of the person producing it; the same enameled scene created by different people can vary between a small masterpiece and an inept daub. The particular technical problems inherent in the enameling process with its need for successive firings, and the fact that colors change during firing at

Colorplate 1.
A chased gilt-bronze jewel cabinet with painted enamel panels: on two hinged doors are female allegorical figures of Autumn and Winter. On the sides (not shown), *The Italian Serenade*, after Watteau, and *Les Charmes de la vie champêtre*, after Boucher. The dial of the agate drum clock is signed: JAMES COX, LONDON. The firm of James Cox of Shoe Lane, London (active c. 1749–72), also made the famous peacock clock that is in the State Hermitage Museum, Leningrad. English, the enamel panels attributed to South Staffordshire, c. 1765. Height: 34.3 cm. (13½ in.). Metropolitan Museum of Art, New York, gift of Irwin Untermyer, 1964

different temperatures, make the production of a really fine enameled piece something to be marveled at.

Attractive though it is as a medium, no artists of consummate genius have appended their names to work in enamel. But paintings and prints by great masters—from Dürer to Braque—have been interpreted by virtuoso enamelers, whose names often are unknown to us. There have been outstanding high points of achievement: the Byzantine era (see colorplate 4); Limoges *champlevé* enamels in the twelfth to thirteenth centuries (see colorplate 17) and painted enamels in the fifteenth to sixteenth centuries (colorplate 2); the Ming dynasty in China (see frontispiece); European seventeenth-century miniatures and eighteenth-century gold boxes (see colorplate 22); English painted enamels in the eighteenth century (see colorplate 25); Japanese nineteenth-century *cloisonné* (see colorplate 12); and a lively if brief Art Nouveau period (see colorplate 26).

Often it is difficult to identify the date and place of origin of an enamel because they were rarely marked. These pieces were highly prized. Sent to and fro as gifts, or seized as booty, they would frequently have been moved from place to place. Furthermore, enamelers themselves traveled widely and worked in courts and workshops far from their original homes, causing an interesting cross-fertilization of styles. Controversy surrounds some of the attributions suggested here; certainly much research remains to be done.

The history of enameling is a fascinating one, full of tantalizing glimpses of vanished cultures. Despite the challenges presented by the techniques, enamelers have rivaled if not surpassed the work of the finest jewelers. In modern times enameling seems to be reverting to its original form, with the solitary craftsperson gaining in importance. But one wonders whether today's artist-enamelers will ever attain the heights of the most splendid work of the past.

Colorplate 2.
A painted enamel-on-copper portrait of Guy Chabot, Baron de Jarnac, by Léonard Limosin (c. 1505–1575/77), who was perhaps the greatest of all Limoges enamelers. Over a thousand enamels of various forms, including many domestic articles, emanated from his workshop. But he is most famous for his portraits, at least 130 of which have survived. French, Limoges, sixteenth century. Height: 23.8 cm. (9⅜ in.). Frick Museum, New York

2 Enameling Techniques

Throughout history, enameling techniques have scarcely changed. The Industrial Revolution may have refined them, introduced greater precision and led to mass production, but the fundamental processes have remained the same since the first enamelers stoked up their furnaces, probably in ancient Greece. Enamel is a form of glass; quite simply, a vitreous glaze that is fused to a metal base. Gold, silver, copper, bronze, iron and steel have been used as bases for enameling. All the enamels described in this book are those applied to a metal foundation.

The metal may be prepared in a number of ways. It can be flat or domed; made into a three-dimensional shape by casting or modeling; patterned by carving, gouging, engraving or soldering; constructed into a box or casket, or fashioned as a hollow vessel. In the case of hollow objects, the metal is beaten or rolled until thin, softened as needed by heating and gradual cooling (a process called annealing), and shaped—by hammering, spinning, jointing or stamping.

Whatever method is used, once the desired form is attained, the metal must be cleansed. In base metals, grease and impurities adhere to the surfaces during shaping. These are removed either by immersing the piece in solvents or by heating it in a furnace to a temperature of approximately 400 degrees Centigrade. The article is then washed and plunged into diluted acid (different acids being used for different metals) to etch the surface in order to give a good key (an allover roughness) to which the enamel can adhere. Finally, the piece must be washed again, rinsed and dried in bran or warm sawdust. Then it is ready to receive the enamel.

Preparation of the enamel is the next stage in the process. Enameling is an unpredictable art—a combination of intuition and science that demands perception as well as skill for a successful conclusion;

Colorplate 3.
An example of the use of composite materials in Celtic enameling. A circular escutcheon from a large bronze hanging bowl, decorated with red cuprite glass and blue *champlevé* enamel inset with small checkered platelets of *millefiori* glass. From the Sutton Hoo Ship Burial, excavated near Woodbridge, Suffolk, England, in 1938-9. Diameter of escutcheon: 6.2 cm. (2½ in.). British, fifth to seventh century A.D. British Museum, London

indeed, it has been compared with expert cookery. The chemical constituents are silica (sand), borates, alkalis (soda and potash), alkalines (lime, magnesia, lead) and oxides of metals for coloring. Silica, the basis for all glasswork, melts and fuses into clear glass when heated with potash or soda and lime. These last impart different qualities to the enamel: potash makes it sparkle and soda gives it elasticity.

There are four different types of enamel: opaque, opalescent, translucent and transparent; but many gradations exist in between. Opalescent and translucent enamels are obtained by mixing opaque and transparent types: opalescent enamel is milky and has a cloudy iridescence. Transparent enamel can be applied as a backing, as decoration, or as a clear glaze over a previously decorated surface to add sparkle and protection to the design. These different types are made by varying the chemical formulae; in other words, by changing the recipe.

The softness or hardness of enamel is an important factor in its use. The harder the enamel, the higher the temperature needed to fuse it. A hard enamel will resist all time's eroding agents, whereas a softer

1.
A stage during the preparation of the constituents of enamel when the hot flux—an almost clear molten glass—is poured onto metal slabs to cool and solidify. The "cakes," or *galettes*, that form are broken up and ground to a fine powder; the enamel is then ready for further processing. This photograph was taken at the Soyer factory at Moulin de Saint-Paul, Condat-sur-Vienne, near Limoges. Soyer has been a manufacturer and supplier to the trade of basic enamel since 1840.

one, although it will be easier to work, can gradually decompose. Borax and lead both have a softening effect; borax also helps the enamel to combine with the various oxides that are the coloring agents. Coloring oxides can be combined with the enamel ingredients when the latter have been heated to form a *flux*—an almost clear molten glass with a slight tinge of blue or green.

After many hours of steady heat, the enamel is either poured onto cold iron slabs, where it cools and solidifies in the form of small cakes approximately 5 inches (13 cm.) across (plate 1), or it can be quenched by being ladled into cold water. At this stage it resembles shattered glass crystals and is known as *frit*.

Color is affected not only by the oxide itself but by the color and nature of the metal base. Blue enamel, for instance, is produced by cobalt. On a silver base the tone will be enhanced, whereas on copper or bronze no intensification in shading is perceived. Carbonate of copper produces green, manganese produces purple, oxide of gold produces some pinks and reds, and so on.

The color is affected by the constituents of the flux as well as by the type or quantity of the oxide. The same oxide can produce different colors when one of the ingredients is changed. If a high proportion of carbonate of soda is added, for example, black oxide of copper produces turquoise blue; but the alternative addition of red lead would produce a yellowish green. The majority of enamel colors cannot be mixed to give an intermediate shade; most of them must be prepared with their own specific oxide.

The next process involves pulverizing the raw enamel with water until it is reduced to a fine powder. The powder is washed approximately six times in distilled water to remove impurities that would cloud and taint the enamel. If it has been ground in a steel mill, the powder must be first washed in nitric acid to dissolve any metal particles, then rinsed repeatedly in water to remove all traces of acid.

Lastly, the enamel powder is dried and sifted through a fine sieve. It is now ready to be applied to the prepared metal surface. Enamel can be applied to metal either in powdered form mixed with distilled water and gum tragacanth or by brush or palette knife, in which case it must first be mixed with a volatile oil such as spike (lavender) oil or oil of sassafras. If a piece is to be covered by being dipped into liquid enamel, or if comparatively large areas are to be sprayed, the enamel is ground with water to form a *slurry*.

Several layers of enamel may be required to form a satisfactory cover, and each must be dried thoroughly, then fired, before the next is applied. Firing takes place in a kiln or furnace at temperatures between 600 and 850 degrees Centigrade (plate 2). The enameled piece is put on a plate, or planche (in earliest times, this would have been made of iron or fireclay; today, a ceramic or stainless-steel

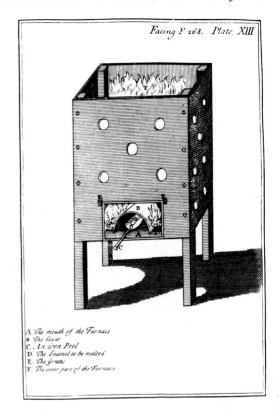

Facing P. 268. Plate. XIII.

A. The mouth of the Furnace
B. The Cover
C. An iron Peel
D. The Enamel to be melted
E. The Grates
F. The inner part of the Furnace

2.
An eighteenth-century enameling kiln illustrated in *Dictionarium Polygraphicum: or, The Whole Body of Arts Regularly Digested,* published in London in 1758. Although today's kilns are powered by electricity or gas instead of solid fuels, the basic principles remain the same. Bantock House Museum, Wolverhampton, England

3.
Cloisonné: enlarged detail from a brush rest for a scholar's table of the Chinese Ch'ienlung period (1736–96). Enamel is built up within the cells of the design and the resulting uneven surface is polished until smooth. Pitting is caused by the bubbling of the enamel during firing.

plate is used), inside a *muffle*—a container used to protect the object from impurities while it is in the furnace. Intense white heat is essential to achieve the temperature at which enamel will fuse to metal.

Firing takes only a few minutes. The piece is then removed and allowed to cool gradually at room temperature. Rapid cooling would have a detrimental effect, creating a brittle enamel that could easily crack or scale. Awkwardly shaped articles that require especially slow cooling are placed on top of the furnace to cool down.

Different colors are fired at varying temperatures, those that can withstand the greatest heat, such as brown, blue and green, being fired first. Certain pinks and turquoises are among the colors that require a lower firing temperature.

The majority of enamel colors are changed by firing, so it is essential for the enameler to be able to visualize the inevitable transformation. Most enameling techniques demand multiple applications and firings, either to fill the cavities in partitioned metal, to cover engraved decoration or to build up a background for painted surfaces. In the case of painted enamels, up to twenty or more firings may be needed to achieve the desired result. Each time an object is fired, great care must be taken to protect already used colors from damage caused by overfiring.

As the expansion of metal is greater than that of enamel, many enameled articles are coated on both sides. If objects made of very thin metal are enameled over the entire surface on one side only, the expansion and contraction of the metal during heating and cooling cause the enamel to crack or craze. Once the metal is enclosed between two layers of enamel, however, the combined substances react simultaneously. Enamel backing is known as *contre-émail (counter-enameling).*

Enameling techniques have evolved over the centuries, but the main forms are extremely ancient. The principal difference between them lies in the way the metal is prepared.

Cloisonné (partitioned) enameling was the earliest method of all. The design is either scribed or indicated by punctures, or it can be drawn in ink onto a smooth metal surface. Cells, or cloisons, to contain the enamel, are formed by placing narrow strips of metal at right angles to the metal base, following the lines and intricacies of the pattern (plate 3). The strips are either soldered into position or kept in place by a suitable adhesive. Enamel colors are placed within the cells and then fired. Opaque enamel is normally used for *cloisonné*, but occasionally other types are also incorporated. When fired, the enamel shrinks, leaving gaps and hollows; these are filled with further applications of enamel, each fired in turn. Finally the surface, which will by now be uneven, is leveled. It is filed down with carborundum until smooth, fired and then polished with the finest pumice powder.

After polishing, the top edges of the metal strips remain exposed, dividing one area of enamel from another and, when newly completed (before the metal tarnishes) lending glitter to the entire surface. The object can then be gilded so that all exposed metal areas retain a thin, untarnishable film of gold.

Champlevé (raised field) enameling followed *cloisonné*. It is the antithesis in technique, as the surface of the metal is gouged away instead of being built up to receive the enamel; it is also a simpler process. Into a thick metal base, which is normally made of bronze or copper, troughs or cells are formed in the metal, either with a burin, by hammering, or by casting, the raised ridges between the cells forming the outlines of the design. Opaque or translucent enamel can be used, but if the base is bronze, its tin content will cause the translucent enamel to become opaque.

The enamel is laid into the excavated spaces; successive applications of enamel and firings follow until the cells are filled and the entire surface can be polished and, if required, gilded. Medieval enamelers from the end of the eleventh to the fourteenth century developed an enriched form of *champlevé* decoration by backing translucent enamels with *paillons*, or foils, of precious metal. Gold had the effect of enriching warm shades of color, whereas silver enhanced cooler tones.

Basse-taille (shallow cut), a refinement on *champlevé*, is a form of enameling in which a pattern or figural composition is engraved or *chased* in low relief into a prepared recess in the metal background, and the entire patterned area is then covered with transparent or translucent enamel. This creates a sense of sculptural modeling due to the varying depths of color, the shading being strongest where the relief is most deeply cut. Gold or silver is normally used for this technique and the reflection of light against the patterned background adds brilliance and beauty to the general effect.

Guilloché (engine-turned) surfaces are used for a type of *basse-taille* enameling. This technique was made possible by the invention in the eighteenth century of the rose-engine lathe (*tour-à-guillocher*), a machine for engine-turning patterns onto metal or other hard substances. Layers of translucent enamel are applied over the decorated metal, creating richly colored, shimmering patterns (plate 4). A fascinating effect can be achieved by varying the colors of the layers: the color of the surface appears to change as the enameled piece is turned this way and that, catching the light at different angles.

Plique-à-jour, which is also known as *émail de plique* (there is no precise translation—it means that daylight shows through), is a type of backless *cloisonné* in which metal cells normally contain transparent enamel. But all types of enamels can be introduced to enhance a design. An openwork pattern is formed with metal strips, or a

4.

Guilloché surfaces: enlarged detail from a Fabergé clock. This type of enameling was made possible by the invention in the mid-eighteenth century of the *tour-à-guillocher*, a machine for engraving metal in regular patterns—a process called engine-turning (*guillochage*). A splendid, luminous effect is achieved when successive layers of different-colored enamels are applied on to a silver or gold *guilloché* background.

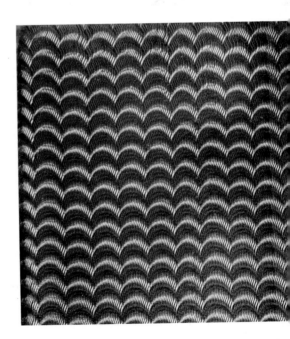

5

6

5.

Plique-à-jour: enlarged detail from the border of a late nineteenth-century Russian bowl. The design is formed in twisted wire and the entire area to be enameled is backed with a substance to which enamel will not adhere. After firing, the backing is removed. The enamel remains suspended within the cells, or cloisons, of the pattern, allowing light to penetrate the translucent areas.

6.

Filigree enamel: enlarged detail from a Fabergé cigarette case. Twisted wire is soldered to a backing plate, enamel is inserted into the cells of the design and fired; painted detail is added before the final firing. Unlike *cloisonné*, the apertures are not filled completely with enamel; nor the surface flattened by polishing.

7.

Filigree enamel of remarkable accomplishment in vivid greens, blue, orange and white on silver, mounted onto a boxwood hand cross. On the reverse (not shown) the boxwood is carved with scenes from the Bible: Christ on the lap of the Father, the Annunciation, the Resurrection, the Baptism and the Ascension. Russian, c. 1700. Height: 19.8 cm. (7¾ in.). Walters Art Gallery, Baltimore

metal object is pierced with holes of various shapes within the area to be decorated, to form the basis of the design. The holes are then filled with different-colored enamels (plate 5).

Two methods are used: in the first, a plate or form to which enamel will not fuse is attached temporarily, to be removed after the enamel has been fired. Among the materials used for this process are mica, fireclay, pumice, tripoli, or rattenstone, and aluminum-bronze. In the alternative method, not employing a backing plate, the enamel is built up gradually by successive applications, each one fired in turn, progressively filling the apertures.

Whichever process is used, care must be taken not to overfire the object since this would liquefy the enamel, making it impossible to maintain its adhesion to the open cells. When the enameling is completed, both sides can be filed until smooth with carborundum, and, finally, polished. Light shining through a *plique-à-jour* enamel gives the effect of a stained-glass window in miniature.

Filigree, or wire, *enameling*, is an extension of the *cloisonné* technique. Here silver or silver-gilt is normally used, but the cells are formed with twisted wire rather than with smooth metal strips. Unlike *cloisonné*, they are not filled to the top with enamel, and the metal wires remain raised instead of being flattened to a smooth surface (plates 6 and 7). The surrounding areas often are decorated with ornate filigree metalwork.

Skan enameling dates back to the sixteenth century in Russia and eastern Europe. It combines the *cloisonné* and filigree methods, using formal floral patterns enclosed by "pearl" borders composed of encrusted drops of white enamel (see plate 37).

Encrusted enameling, known as *émail en ronde bosse* (enamel in rounded relief; see colorplate 16), was used to decorate irregular, high-relief surfaces as well as pieces modeled in the round—small-scale sculptures. This method occasionally is also referred to as *émail en blanc* (enamel in white), which is confusing since it implies, incorrectly, that the total decoration is white; the term came into vogue because of the frequent use of white, particularly for faces and limbs. Opaque or translucent enamel can be used, but this technique is extremely laborious. The enamel is required to cover three-dimensional forms, which means that the process of ensuring the adhesion to metal (normally gold) of a vitreous matter that melts during firing is an exceptionally hazardous one. One method is to use plaster of paris to cover parts that are not to be decorated, thereby supporting the piece during firing and leaving exposed the enameled areas.

Enameling *en plein* (on an open field) is a most difficult technique to accomplish. Comparatively large areas of enamel are "floated" onto gold or silver surfaces, creating an additional layer rather than filling prepared recesses (plate 8). The method was first developed in Paris in the mid-eighteenth century, and many of the finest gold boxes of the period were enameled *en plein*.

Enameling *en résille sur verre* (in grooves on glass) is a rare technique that was in use for only sixty to seventy years, from 1570–80 to about 1640. A design is cut into glass, normally blue, turquoise or green. The incisions are lined with gold foil, then filled with colored enamels of comparatively low firing temperature; the thin gold remains visible at the top edges, thereby outlining the enamel. This delicate process was suitable only for decorating small articles such as medallions and watchcases.

7

8.
En plein enameling dates back to the eighteenth century. Although it is normally associated with elaborately painted French gold boxes, this beautiful silver jug, enameled in black, demonstrates its effectiveness when used in the modern idiom. Made by Professor Gerald Benney of the Royal College of Art, London, who revived the *en plein* technique in 1969. Height: 20.3 cm. (7⅞ in.). Commissioned in 1978 by the Victoria and Albert Museum, London

Millefiori "enamel": from the first to approximately the eighth century *champlevé* objects occasionally were inset with *millefiori* pieces (colorplate 3). In the first to early third century some bronze objects were partly covered with so-called *millefiori* enamel (see plate 14). This was not true enamel but slivers of *millefiori* glass that were placed on a metal base and heated just sufficiently to adhere to the metal and each other without being melted.

Painted enamels In each of the enameling techniques described above, metal was a visible partner with enamel in the finished work. But no matter how varied the colors or methods employed, the expression of enameled pictorial art lacked realism. This situation was remedied in the third decade of the fifteenth century when artists began to paint with enamels. The restrictions inevitably imposed by the role of metal were removed when painted enamels evolved; yet something was lost, too. For approximately two thousand years the marriage between glass and metal had endowed enamels with a specific, individual quality. But once the metal served only as a backing to the painted work, fulfilling a purpose similar to that of the canvas in an oil painting, this special quality was lost.

An early rare example known as the Ara Coeli medallion, painted in monochrome white and gold on a dark blue ground, dates from about 1425 (see plate 59). And a plaque dated about 1450 attributed to the painter Jean Fouquet displays a technique known as *enlevage à l'aiguille*. Here a black, or very dark, background is covered with a layer of gold or white enamel, which is then scratched with a needle to reveal the underlying color.

By 1470 painted enamels were being produced at Limoges in central France, where several methods were employed. In one, black or a very dark color again formed the background; a layer of white opaque enamel was applied, and the design drawn on this with a sharp point. After firing, the dark background was revealed through the lines, the finished work resembling an engraving. Brushwork was added to the white surface, at first with dark shades, then later with brighter colors.

Around 1535 enameling *en grisaille* (in gray tones) was introduced. White opaque enamel was applied to details of a picture in varying thicknesses over a dark background, resulting in shading from gray tones to white, the stronger white areas being used to model highlights on figural compositions. Colors were then added and sometimes gold or silver was applied, either as *paillons* (small pieces of foil) or in thin lines on hatched areas, to emphasize detail. Another effect, called *joailleries*, was to apply tiny drops of enamel onto foil to simulate jewelry.

Around 1630 small gold objects covered with white enamel and finely painted in a full spectrum of colors began to be made, an innovation attributed to Jean Toutin (1578–1644), a goldsmith from Châteaudun, near Orléans. This was the beginning of the great era of miniature painting on enamel. Throughout the seventeenth and eighteenth centuries in both Europe and Great Britain exquisite portraits, watchcases, snuffboxes and other personal accessories were decorated in this manner.

During the mid-eighteenth century enameling on copper became a highly successful industry, especially in England, where the decoration was either hand-painted or applied by *transfer printing*, an English invention. In transfer printing (see plates 81, 82 and 86), a design is engraved on a copper plate and liquid enamel is rubbed over it. The plate is polished, leaving enamel only in the incised lines of the engraving. Thin gummed paper is then placed onto the plate, covered with flannel, and the "sandwich" is rolled through a press under considerable pressure. When the paper is removed, the design is seen to have adhered to it in minute ridges of color. The enamel object is smeared with linseed oil, the paper is smoothed onto it and rubbed all over to ensure complete contact. Next, the piece is fired and the paper burns away, leaving the design imprinted on the enamel surface.

There were other methods of transfer printing but the one described here was the most generally used. Finely detailed designs were frequently used in monochrome, or very lightly colored with translucent enamels; but others were intended to be used solely as an outline guide for painting fully with opaque colors.

All the processes described above present certain technical hazards. On close scrutiny few enamels, especially those employing early methods, are devoid of blemishes: pitting, irregularity of outline, minuscule cracks on the surface. But these minor imperfections are natural to the medium and do not detract from the interest or value of pieces that, over the centuries, have often retained their color and brilliance undiminished.

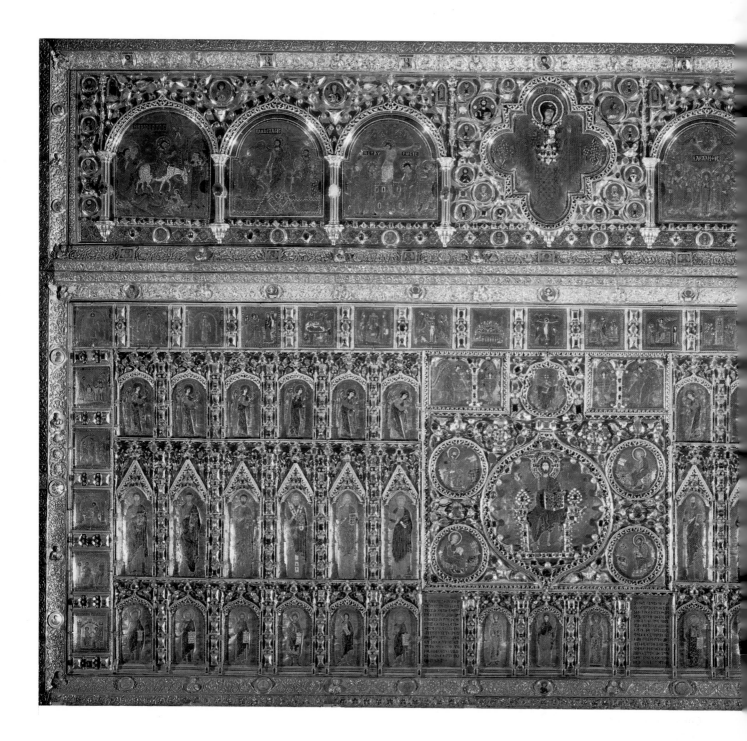

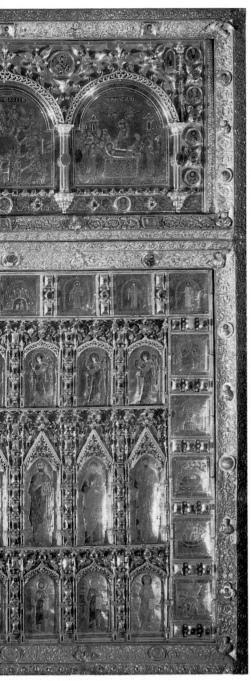

Colorplate 4.
Brought from Constantinople to Venice, the Pala d'Oro was erected in 1115 by the doge Ordelaffo Falier, and renewed in 1209 and 1345 according to a fourteenth-century inscription. The altarpiece incorporates 137 *cloisonné* enamel tableaux and portraits set on a gold background, surrounded by borders of precious stones and Gothic framework. The centerpiece is a picture of Christ enthroned, surrounded by the four Evangelists. Byzantine, tenth to twelfth century. Width: 29.21 cm. (11 ft., 6 in.). St. Mark's Cathedral, Venice

3 Byzantium and Before

The search for early examples of enameling is bedeviled by many problems. When tombs were plundered, fine enamelwork may well have been seized, and in many cases it would have been melted down for its gold base. Time itself can be a formidable enemy of early enamels, which are exceedingly fragile; the chances of delicate *cloisonné* work surviving for millennia are obviously slim.

Often civilizations appear to have evolved art forms close to enameling but they lacked the fundamental principle of fusing molten glass to metal. Around the eighteenth century B.C. the Egyptians pioneered a prototypical form of the *cloisonné* technique (plate 9). Elaborate gold ornaments were inlaid with pieces of precious stones, which were held in place by ribs soldered onto a metal base, normally gold. A pectoral jewel found on the mummy of the young Tutankhamun (reigned c. 1334–1325 B.C.) had cloisons filled with red and lapis lazuli–colored glass that appear to have been inserted in powder form and then fired.

The earliest genuine examples of enameling are considered to be Mycenaean, excavated from graves on the mainland of Greece, in Crete and Cyprus, and dating between the thirteenth and the eleventh centuries B.C. In his book *Chinese and Japanese Cloisonné Enamels*, Sir Harry Garner writes about a unique gold scepter: "This remarkable scepter, which appears to be in true *cloisonné* enamel, reaches a standard of technical achievement which was not to be met again for nearly 2,000 years." Enameling may have been introduced into Mycenaean Greece from Asia Minor, since Asiatic influence is present in much Mycenaean decoration.

The next true enamels are found again in Greece. In the sixth to third century B.C. Greek gold jewelry in wire openwork (filigree)

9

10

11

9.
Among the earliest known examples of the *cloisonné* technique is an Egyptian winged scarab, made of electrum (an alloy of gold and silver) inlaid with carnelian, green felspar and lapis lazuli. The piece dates back to the eighteenth century B.C. It was not until some five centuries later that jewelers in countries bordering the eastern Mediterranean attempted to insert and fire glass into cloisons formed in precious metals. Egyptian, Twelfth dynasty, Sesostris II, about 1885 B.C. Height: 1.8 cm. (¾ in.). British Museum, London

10.
This bronze helmet, embellished with iron bands, gold, and red enamel, or cuprite glass, was found in 1861 in an old streambed of the Seine at Amfreville-sous-les-Monts, Eure. Although it was discovered in a northern area, the helmet probably was a trophy of Italian campaigns or the work of a craftsman trained in the transalpine region. European Iron Age, late fourth century B.C. Height: 17.5 cm. (6⅞ in.). Musée des Antiquités Nationales, Saint-Germain-en-Laye, France

11.
A relic of the British Iron Age, this shield was recovered from the River Thames at Battersea in 1857. The substance its twenty-five roundels contained was long considered to be red enamel, until research at the British Museum revealed it to be cuprite glass, the sealing-wax red being caused by crystals of cuprite oxide in its composition. First century A.D. Length: 84.5 cm. (33 in.). British Museum, London

appears, but the blue and white enamel detail is minute and appears only incidental to the central design impulse.

Enamelwork attributed to periods from the ninth century B.C. onward has been found in Caucasian, Egyptian and Etruscan tombs, and some connection is thought to exist between them. But even in pre-Christian times enamels were probably exported widely; consequently, their place of origin is uncertain.

In Iran, enameling on terra cotta was produced in the fifteenth century B.C. The Oxus Treasure—which dates from the sixth to fourth century B.C. and is now in the British Museum, London—includes elaborate gold jewelry decorated by the *cloisonné* technique, but the inlay, most of which is missing, was of precious stones and paste, not enamel. It seems that the earliest authenticated examples of Persian enamelwork are from the Sassanian period (A.D. 224–642; see plate 19).

The earliest Asiatic enamels, found in graves dating from the fifth century B.C., show marked Hellenistic influence. But as time passed, the vigorous and distinctive Asiatic polychromatic tradition asserted itself, and in the late fourth century B.C. the Sarmatians came on the scene.

The Sarmatians were a Persian nomadic tribe from the area north of Lake Aral, east of the Caspian Sea, who spread throughout southern Russia and far beyond. Their work is highly distinctive, being richly decorated with wild animals, flowers and geometric ornament, including small amounts of enamel inlaid in gold and silver.

Some scholars suggest that from the second century B.C. the Sarmatians were responsible for spreading the art of enameling throughout Europe to Spain, North Africa and even Britain. But finds at La Tène in Switzerland, and elsewhere, indicate that western European enamelwork might date from as early as the fifth century B.C.; that is, before the Sarmatians had reached the West (plate 10). In any case, in the northern provinces of the Roman Empire craftsmanship flourished. The Celts in Belgium, France, England, Scotland and Ireland were adept at enameling (plate 11).

A Greek philosopher, Philostratus (A.D. 170–245), traveled to Gaul and Britain in 208. He described a boar hunt with riders whose mounts had horse trappings decorated in vivid colors. Philostratus possibly was referring to the artistic activities of the English or Irish when he wrote: "It is said that the barbarians who live in the ocean pour these colors into bronze molds, so that the colors become hard as stone, preserving the designs."

Enameled horse trappings have been discovered in many places in the British Isles (colorplate 5). Philostratus's elementary description of *champlevé* indicates that such work was unknown at the time in Greece or at the court of the empress Julia in Rome, where he lived.

12.

The Birdlip mirror dates from immediately before the Roman conquest of Britain in A.D. 43. It was discovered in 1879 by a workman quarrying for stone at an early Iron Age site east of the city of Gloucester. Made of bronze, the back of the mirror is engraved in a curvilinear pattern, inset with studs of red enamel. British, early first century A.D. Length: 38.7 cm. (15⅛ in.). Gloucester City Museum and Art Gallery, England

13.

A massive pair of armlets of the North British Iron Age, found in Aberdeenshire, Scotland. Cast in bronze as complete hoops, they are finely polished, with red and yellow *champlevé* enamel in a checkered pattern in the central disks. First to second century A.D. Diameter: approx. 14 cm. (5½ in.). British Museum, London

12

13

The Celts produced enameled bronze objects and jewelry (plates 12 and 13). By the first century A.D. they also were masters of the Roman *millefiori* technique, and some bronze objects had extensive areas covered in so-called *millefiori* enamel (plate 14). The remains of a Roman villa, now called Anthée, near Dinant in Belgium, suggest that an enameling atelier manufacturing such wares was active there, probably between the second and early third centuries A.D. Sticks of *millefiori* have been found on the sites of two Irish workshops, at Garranes dating from the fifth or sixth century, and at Lagore from the seventh or eighth century.

Colorplate 5.
A bronze bridle-bit found at Rise, near
Hull, in Yorkshire. The cheek-pieces are
decorated with *champlevé* enamel in red and
blue, and the matte surface of the bronze,
which is in very good condition, is flecked
with gold. Length: 27.9 cm. (11 in.). English or Irish, first century A.D. British
Museum, London

14

Larger objects made by Celtic enamelers include vases, skillets, buckets and plaques, some of them enameled over almost the entire surface with bands or panels of multicolored decoration (plate 15).

It might be thought that from such splendid, if scattered, beginnings, enameling as a coherent and continuously developing art form would have gathered momentum gradually during the first few centuries after Christ. But a disastrous period of wars, invasions, famines and plagues played havoc with artistic enterprise. As entire populations reeled from these successive blows, the struggle for mere sur-

15

vival came first. Here and there a relic of enameling appears, but these early centuries—the Dark Ages—provide few examples, until new glories emerged at Byzantium.

14.
A hexagonal bronze pyx, decorated on the top and sides with so-called *millefiori* enamel in red, white, and blue with some green patination. This was not true enamel but many small, cross-section pieces of *millefiori* glass, adhered to each other and to the bronze foundation. Provincial Roman, possibly made in the northern part of Gaul, second century to first half of the third century A.D. Height: 6.1 cm. (2⅜ in.). Metropolitan Museum of Art, New York, Fletcher Fund, 1947

15.
It was reported that 1,500 third-century coins, the latest being those of Tetricus, governor of Aquitaine (268–73), were with this vase when it was found in the nineteenth century at La Guierche, a village thirty-five miles west of Limoges. Decorated with a series of vertical bands composed of facing pairs of trumpets and hearts, the *champlevé* enamel is dark blue, red, olive green and turquoise. Provincial Roman, third century A.D. Height: 12.1 cm. (4¾ in.). Metropolitan Museum of Art, New York, Fletcher Fund, 1947

Byzantium The golden age of Byzantine art, from the sixth to the beginning of the thirteenth century, was the most significant period of development in the history of enameling. The initial dynamic thrust of Christianity transformed all the arts. Not only was the subject matter of art revolutionized, concentrating on Christ's life and episodes from the Bible, but so too was the rapidly developing Church, whose rituals required large numbers of liturgical objects. Byzantine workshops were sufficiently advanced to decorate these highly prized artifacts with enamel.

Byzantine art developed in the eastern part of the Roman Empire after Constantine (reigned 306–37), the first Christian emperor, chose the site of the ancient Greek city of Byzantium as his new capital and in A.D. 330 named it Constantinople.

The era of Justinian I (reigned 527–65) was richly productive in the arts. Among the first recorded enamels of this period were those on Justinian's golden altar in the great church of Hagia Sophia in Constantinople. But the Persians and the Arabs both challenged Byzantium, and a long series of wars and campaigns during the seventh century brought artistic production to a low ebb.

An internal struggle, the Iconoclastic dispute, which began in 726, also had a vital effect on the progress of the arts. The Iconoclasts condemned, and where possible destroyed, religious figural representation. They were not finally overthrown until 843.

Political and religious organization, trade and dynastic marriage spread the influence of Byzantium throughout Europe. Gifts—sometimes church ornaments, sometimes crowns—were often sent by the reigning emperor to his fellow princes and potentates in Europe. Pilgrims and crusaders carried away valuable objects. But the most dramatic way in which Byzantine art reached the West was through the brutal sacking of Constantinople in 1204.

The Fourth Crusade (1202–4) had set out to liberate the tomb of Christ from Muslim hold. Instead, incited by the doge of Venice, the crusaders destroyed the Christian Empire in the East and ransacked Constantinople. The city's incalculable riches, relics and works of art were seized and carried home in triumph to western Europe, where the impudent looters generously presented the spoils to their local abbeys and churches.

The enduring irony is that by such disgraceful behavior they did in fact ensure the preservation of much Byzantine art. Western churches offered far greater protection than those in the East. When Constantinople fell to the Turks under Muhammad II in 1453, its remaining precious works of art simply disappeared forever (many were probably melted down). Enameling declined from this time, and the craft was relegated to the decoration of inexpensive devotional objects.

Most Byzantine enameling was *cloisonné* on gold or electrum, an alloy of gold and silver, the early work being decorated with abstract and geometric symbols.

Around the year 700 the first figural elements appear: human figures are shown, but these early representations are understandably primitive. The figures are drawn out of proportion, thickset, with big, doll-like heads. Few examples survive from this period, and the Iconoclast controversy caused a hiatus before the main Byzantine development began in the mid-ninth century.

Two methods of *cloisonné* enameling were employed: in the one more generally used, the entire surface of an area was covered with cells that were filled with enamel. In the alternative process, a recess

16.
A gold pendant employing the *cloisonné* technique. A recess is formed in a gold plate within the outlines of a design and cloisons are placed within the hollow and filled with enamels; finally the surface is polished so that the enamel is level with the surrounding gold background, resulting in a flush surface. Russo-Byzantine, twelfth century. Diameter: 4.7 cm. (1⅞ in.). Walters Art Gallery, Baltimore

17

was formed in the metal and the cells were built up within it. When filled with enamel and polished, the enameled area rose to the level of the surrounding field of gold, thus creating a flush surface (plate 16).

Reliquaries, book covers and jewels were made; enameled medallions and plaques were sometimes incorporated into crowns. Later, human figures became more elongated, the faces oval, with features and draperies suggested by a few well-placed lines. With growing experience and confidence, the enamelers gained precision and judgment (plates 17 and 18).

The most remarkable example of Byzantine enamelwork is without doubt the Pala d'Oro in St. Mark's, Venice (colorplate 4), a triumphant altarpiece conceived on an epic scale for enamels. Scholars are still locked in debate about its composition, history and provenance; but the authority and richness of this work sum up the achievement of Byzantine enameling.

18

17 and 18.
Medallions portraying the Virgin and St. John the Baptist as intercessors with Christ for the salvation of men's souls. These enamels exemplify the finest Byzantine gold *cloisonné* work: the expressions and gestures are conveyed with such economy of line that they might have been drawn free hand, belying the complexities of the processes involved. These and nine other medallions portraying busts of saints were a part of a large icon of St. Gabriel, now no longer in existence. Russo-Byzantine, eleventh century. Diameter: 8.3 cm. (3¼ in.). Metropolitan Museum of Art, New York, gift of J. Pierpont Morgan

4 Persia and India

Persia   Apart from Byzantium, Iran is the only Middle Eastern area whose enameling is distinctive and important. Predictably, the luxury-loving Persians were attracted by enamels, and their work dates from a very early period, that of the Sassanians (A.D. 224–642). A clasp found at Risano in Dalmatia (plate 19) is in superior *cloisonné* enamel. It has a motif composed of four hearts forming a rosette; the same design appears in other Sassanian work.

A most important twelfth-century *cloisonné* copper bowl known as the Ortokid dish (plate 20) is the subject of much conjecture, for it bears inscriptions in imperfect Persian and Arabic, and is thought to be the work of a foreign, possibly Byzantine, artisan.

Timur the Lame (c. 1336–1405), although a ruthless barbarian, encouraged the arts and is reputed to have had a passion for elaborate ornamentation, including enamels. The doors of his own reception room were of blue-enameled silver with gold insets. This, too, could well have been late Byzantine work, looted during one of his campaigns.

Fourteenth- and fifteenth-century Hispano-Moresque swords had *cloisonné* enamel on their handles and mounts. The Safavid period (1499–1736) was a great age of Persian art generally, especially under Shah Abbas the Great (reigned 1587–1629), yet, strangely, no examples of Safavid enamels survive in Iran. At the Royal Treasury in Wawel Castle, Cracow, however, there is an early seventeenth-century enameled mace with a pear-shaped head, called a *bulawa*, that was made in Persia or by Persian craftsmen (plate 21).

A wealth of Mughal enamels was taken to Persia in 1739, when Nādir Shah (reigned 1736–47) invaded India and sacked the treasury at Delhi. Among the booty (which included the Kohinoor diamond) was the fabulous jewel-encrusted Peacock Throne, made for Shah

Colorplate 6.
An outstanding example of Qajar enameling by one of the Fath Ali Shah's chief enamel painters, Muhammad Ja'far: a circular gold dish, presented on behalf of the king by his ambassador, Abul-Hasan Khan, to the directors of the East India Company on June 18, 1819. The border is decorated with painted enamel flowers and birds. In the center, surrounded by flower bouquets within scrolled panels, is the Lion and Sun. An inscription on the back of the dish records the presentation. Persian, dated 1233 (1818). Diameter: 26.7 cm. (10½ in.). Victoria and Albert Museum, London

19.
A *cloisonné* enamel clasp excavated in 1978 at Carina, Risano, in Dalmatia. It was acquired by the renowned archaeologist Arthur Evans in the Balkans. Part of a composite necklace, the design on one side—a rosette composed of four hearts with four alternating pointed leaves—is similar to motifs found in late Sassanian ruins. Sassanian or Byzantine, sixth century. Diameter: 1.8 cm. (¾ in.). Ashmolean Museum, Oxford

Jahan of India (reigned 1628–58). The throne and many other treasures were broken up following the assassination of Nādir Shah by the Kurds in 1747. Around 1800 another throne, known today as the Peacock Throne, was made in Persia for Fath Ali Shah (reigned 1797–1835). It was used on the occasion of his marriage in 1809 to an Isfahani lady nicknamed the Peacock Lady. This throne is richly enameled and set with jewels, and it is thought that some of the jewels might have come from the original Peacock Throne.

From the eighteenth century in Persia and Turkey *champlevé* and painted enamel, both techniques often being combined in the decoration of a single piece, were used on sheaths and handles of daggers, as well as on domestic articles such as water pipes. A type of encrusted enameling also was practiced, in which raised globules of enamel were used to simulate precious stones (plate 22).

The greatest center of Persian enameling, which is known as *minakari*, was at Bibihan, near Shiraz, although there has also been a continuous tradition at Isfahan from the Safavid period right up to the twentieth century. In the late eighteenth century the distinctive style of painting in enamels known as Qajar first appeared. Fath Ali Shah was a lavish patron of the arts; his portrait appears on no less than nine enamels among the crown jewels.

Colorplate 7.
Paintings of Persian beauties, usually depicted enjoying leisurely pursuits, were favorite subjects on Qajar enamels. These portraits appear on two bowls from *qualians* (water pipes). The taller is enameled on gilt copper with part of the original stem of the *qualian* still attached; the smaller, enameled on gold, has been mounted on a silver base of later date, to form a cup. Persian, early nineteenth century. Height of taller *qualian*: 15.5 cm. (6 in.). Victoria and Albert Museum, London

22.
A small coffeepot encrusted with globules of turquoise-colored raised enamel, simulating turquoise stones, on a gilt-copper foundation. Persian, eighteenth century. Height: 13 cm. (5⅛ in.). Victoria and Albert Museum, London

The detail of Qajar painting on enameled gold or silver was superb: sumptuous brocades and jewels were miraculously conjured up. Portrait medallions (colorplate 7) were sometimes surrounded with borders of pink cabbage roses and fine raised gold enamel on a royal blue background. Birds, flowers and figures adorned many beautiful objects—the whole effect was ostentatious but exceedingly charming.

The most distinguished painters in the medium were Agha Ali, a meticulous and elegant craftsman, son of the artist Agha Bagher; and Muhammad Ja'far, whose work was somewhat stiffer and on a larger scale. Ja'far made a huge enameled gold dish that was presented to the directors of the East India Company (colorplate 6).

After the mid-1830s no superior work was produced, although in the 1860s Kazim ibn Najaf Ali established himself as a preeminent artist in enamels. He was a versatile painter, excelling in lacquerwork as well as enameling, but his standards deteriorated. The figural designs became increasingly dependent on European models, which were in themselves derivative and dull.

The aesthetic value of these objects varies, yet the painted scenes that decorate water pipes, vases, candlesticks and dishes give us a glimpse of a way of life that has now vanished forever. This is one of the incidental benefits of the art of enameling, for such glimpses preserve fragments of lost worlds.

Enameling revived in Persia after 1931, and until the 1970s was dominated by one man, Shokrollah Saniezadeh, a gifted painter of miniatures. In 1977, about five hundred skilled workers in Isfahan were engaged in making everything from enamel earrings to large wall hangings. Doors and windows for sacred mausoleums and other holy shrines have been commissioned from Isfahan craftsmen but modern work lacks the jewel-like quality of the older enamels.

India Only isolated examples survive of early work considered to be Indian: a pendant of the fifth century, for instance, which was found at Sirkap (Taxila) and is now in the Cleveland Museum of Art. For centuries the eyes of bronze buddhas were enameled, but no major extant enamelwork dates from earlier than the sixteenth century. Perhaps, during the early fifteenth century, enameling reached India from the West by way of the Silk Route, through western Asia; another theory is that the Mughals from Persia were responsible for its introduction. But no evidence survives to support either belief.

The Mughals were the Muslim rulers of India, who were in power from 1526 until 1858. Most important were Humayun (reigned 1530–56), who spent the years 1540 to 1544 in exile in Persia, and his son Akbar (reigned 1556–1605). Humayun brought an enthusiasm for Persian culture back with him to India, and Akbar became a great patron of the arts. During his reign the characteristic Mughal style developed, consisting mainly of patterns of leaves and flowers.

Under Shah Jahan, the builder of the Taj Mahal, a more Islamic note crept into artistic design, and miniature painting and metalwork flourished.

Enameling was practiced mainly in the northwest part of India, the original center being the Punjab, and it was from here that the craft spread to other areas. Lahore, the capital of the Punjab, produced some of the finest work, and the city has always been a principal source of raw materials. Punjabi craftsmen and their descendants worked at *minakari* (enameling) in all the major centers, including Jaipur, Delhi, Lucknow and Benares.

Indian enamels have a richness and brilliance that is unique: the vivid colors, combined with gleaming gold; the appealing figures, animals and flowers; the lavish enrichment of precious stones—selected in most cases for their color rather than their value. The exotic, incandescent effect is such that it is no surprise to learn from Momin Latif, an authority on the Mughal period in India, that "Indian craftsmen work in a state of almost mystical ecstasy. The tools are treated with respect and are even consecrated on Holy days at the appropriate temples . . . the great secrecy which craftsmen everywhere attach to their work is in India not just commercial but part of a mystery."

The *sonar*, or goldsmith, would prepare the gold work and then the *minakar*, or enameler, and his family would attend to the actual enameling processes (colorplates 8 and 9). The most favored technique in India is *champlevé*, in which ridges or flat areas of gold surround islands of brilliant color; the *basse-taille* method is also employed, in which the engraved gold shines through translucent colors; painted enamels are highly esteemed but *cloisonné* is rarely used. The vast majority of Indian work is on gold or silver, but from the seventeenth century or earlier, at Multan in the Punjab, decorative and domestic objects were enameled in turquoise blue on a base of copper. Experts differ regarding the origins of enameling in Jaipur (formerly called Amber), the capital of Rajasthan. Some propose that an enameling manufactory was established there by Maharajah Man Singh during the third quarter of the sixteenth century. Others insist that the early 1700s was more likely. But, whichever the case, enameling flourished in Jaipur (plates 23, 24 and 25), practiced by Sikh families. In 1884 there were nine *minakars*, all master craftsmen; today there is but one: Kudrat Singh. Jaipur enamels cover a wide range of objects, from jewelry such as amulets to camel and elephant trappings. A favorite was a mango-shaped locket used by Hindus to contain scent and by Muhammadans to house a little compass that indicated the direction to Mecca.

In the seventeenth century, in northwest India, enameling was used to decorate the handles and hilts of daggers and swords. Examples are on display today in the National Museum, Delhi. Another martial use of the craft sprang up in the following century at Jaipur, at Delhi

23.
A gold perfume locket, probably made in Jaipur. *Cloisonné* enamel red flowers and turquoise leaves flank the figure of Nayika, a heroine pictured in amorous mood awaiting her lover, wearing a red shawl over her engraved gold sari, and set against a blue background. On the reverse, two dancers are depicted against a red background. Such themes often were used in miniature paintings and the decorative art of the period. Indian, Rajasthan, c. 1700. Width 3.5 cm. (1⅜ in.). Cleveland Museum of Art

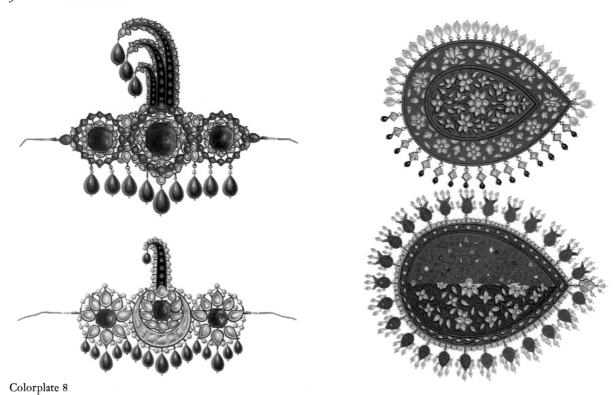

Colorplate 8

Colorplate 9

Colorplate 8 and Colorplate 9.
Illustrations by "Ram Bux, son of Esur" from a book written in the late nineteenth century by two English Army officers serving in India. They explain that the enameler works in his house assisted by most of his family and that ". . . rare old designs, valuable ornaments, and pigments are tied up in dirty rags, and kept in niches in the walls of the rooms."
Below: a group of enamelers engaged in activities prior to the firing of enamels; *above right:* designs for epaulets; *above left:* designs for forehead ornaments. From *Jeypore Enamels*, by Lieutenant-Colonel S. S. Jacob and Surgeon-Major T. H. Hendley (London: W. Griggs, 1886)

24.
A gold pendant, *champlevé* enameled in opaque white, blue and lavender, and translucent red and green. The subject is similar to those found in contemporary Rajasthani paintings: Krishna, the Divine Herdsman, attended by two cowherds bearing offerings. Indian, Rajasthan, late sixteenth century. Height: 5.1 cm. (2 in.). Cleveland Museum of Art

25.
Almost certainly made for a royal patron by Hindu goldsmiths and Sikh craftsmen, this gold pendant is decorated in *basse-taille*. Engraved gold gleams through translucent blue, green and red enamel to illustrate the *Ten Avatars of Visnu*, depicting the incarnation of a divine person. The central panel represents the dialogue of Krishna with Arujunda the warrior, seated in his battle chariot. The reverse is studded with brilliantly colored precious stones. Indian, Rajasthan, late sixteenth century. Height: 4.9 cm. (1⅞ in.). Museum of Fine Arts, Boston

and at Hyderabad in central India, where archery equipment was decorated with enameling on parts of the metal bows, and on the protective finger ring, or *mudrika*, worn by the archer.

From the eighteenth century, Benares, in the north of the country, became a famous center for painted enamels on a white ground, introduced there from Isfahan in Persia. Benares enamels are distinguished by the predominance of pink coloring, and a specialty of work from the city is the combined use of *champlevé* together with painted decoration. This most attractive type of enameling is still practiced in Benares; some striking examples exist on gold jewelry and ornaments.

In the eighteenth century a distinctive style developed in Lucknow consisting of green and blue *champlevé* enamel on silver. This is also considered to have derived from Persia, although the Lucknow enamels are superior in quality. In Kashmir, *champlevé* enameling on silver was more successful than on gold, and objects were given an allover pattern that consisted of dark green or delicate light blue, some pale turquoise blue and brownish red.

24

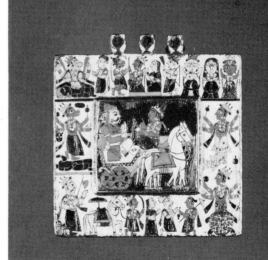

25

5   The Far East

China   Exactly how enameling was introduced into China is a vexed question. One theory holds that as the Mongols thrust westward during the thirteenth century, Byzantine and Persian ideas and Western artistic influences were fed back into China, among them, examples of enameling. The Chinese term for enamel is *fa lan* or *fo-lang*, from the Chinese name for Byzantium; enamels are also referred to as *ta shih yao*, or Arabian ware.

It has been suggested that enamels were produced during the T'ang dynasty (618–906). In the Shoso-in, the Imperial Treasury at Nara in Japan, there is a silver mirror with elaborate *cloisonné* decoration on the back that was long considered a unique example of Chinese enamelwork of that period. Originally it was attributed to eighth-century Japan, but in fact such sophisticated enameling was unlikely to have started in the Far East at that time and then lain dormant for more than five centuries. The mirror is now considered to be Japanese, from the seventeenth century or later.

The next period is said to have been during the Yüan dynasty (1280–1368), the era of the Mongolian rulers. Again it is possible that enameled articles found in China, and thought to date from this dynasty or before, might have originated elsewhere—perhaps in Byzantium. The famous Ortokid copper dish (see plate 20), thought to have been made in Persia or Byzantium between 1114 and 1144, is so similar in technique and design to early Chinese enamels that a close relationship between them is indicated. Dating early Chinese enamels is in any case a hazardous undertaking, for they rarely carry reliable marks or inscriptions revealing date or dynasty, although such practice was common in the case of pottery, lacquer and stone carvings.

Colorplate 10.
Fish and other water animals swim among water plants on a turquoise blue ground inside this large *cloisonné* enamel bronze bowl. The exterior is decorated with the theme of "The Hundred Deer" and the base is enameled with random plum blossom. Chinese, Ch'ien-lung period (1736–96). Diameter: 62.2 cm. (24¼ in.). The Avery Brundage Collection, San Francisco

Some of the earliest Chinese enameling was *champlevé* on bronze. Since the Chinese had already been making carved bronze objects for centuries, the evolution of *champlevé* seems entirely natural. The difficulty of carving in bronze kept the early designs simple; they are bold, striking and often dramatic, contrasting with more decorated later work.

Although the Chinese themselves claim that the earliest *cloisonné* enamels were produced during the Yüan dynasty, no known work can confidently be identified as belonging to such an early period. The earliest suggested date is the beginning of the fifteenth century. During the reign of Hsüan Te (1426–36) of the Ming dynasty, *cloisonné* was featured on ritual cast-bronze vessels; a few surviving pieces bear the mark of this period.

The majority of Chinese designs derive from the natural world. First came stylized plants; then with increasing frequency these were combined with animals, especially the peculiarly Chinese bestiary of mythological creatures with their attendant magical powers: dragons, phoenixes, ogres, lions. Flowers, too, had their symbolic significance, each month of the year sporting its own emblem, such as the June pomegranate and the December poppy. Landscapes were often featured, but not until the second half of the sixteenth century did human figures appear in *cloisonné* work.

Early Chinese *cloisonné* was carried out on heavy cast-bronze vessels, with thin bronze strips forming the cells. The strips were hammered, not drawn (that is, they were flat, not rounded), and of uneven thickness, as becomes apparent when the polished surface of an early piece is scrutinized. During the sixteenth century, vessels appeared that were obviously formed by hammering from sheet metal, which had a higher copper content to render it malleable. This produced a lighter, thinner body that had reasonable strength (plate 26). In the seventeenth century drawn copper wires generally began to be used for *cloisonné* work, and these were of even thickness throughout the decoration of the object (plates 27 and 28).

Almost all superior pieces of Chinese *cloisonné* were gilded, but this is not always apparent because the exposed gilding on rims and mounts suffers through wear. Silver was occasionally used for Chinese *cloisonné*, *basse-taille* and *repoussé* work, but gold appears only rarely. Different techniques were sometimes combined in a single piece—for example, *champlevé* for the bolder elements of a design, surrounding finer *cloisonné* work.

Although the Chinese considered enameling a comparatively minor art, enamel objects of great interest were made there—from exquisite small figural pieces to magnificent large bowls and ornaments (plate 29; see also frontispiece).

Up to the mid-fifteenth century, Chinese enamels are vivid, plain and simple in color. The shades range from turquoise blue through

26.
A large *cloisonné* ewer following an ancient Persian form, used as a rose-water sprinkler. It is decorated with lotus blossoms and leafy scrolls in multicolors on a turquoise ground, and a circular panel on each side encloses fruit and flower motifs (emblems of abundance), with the pomegranate and the citron, known as "Buddha's hand print," predominating. Height: 33.7 cm. (13⅛ in.). Chinese, sixteenth to seventeenth century. Brooklyn Museum, gift of Samuel P. Avery

27.
A *cloisonné* enamel garden seat in the shape of a temple drum, decorated with flying phoenixes and the flowers of the four seasons in multicolored enamels on a turquoise ground. On the top panel there is a design of lions playing with a brocaded ball, and Buddhistic symbols. Chinese, K'ang-hsi period (1662–1722). Height: 39.4 cm. (15⅜ in.). Brooklyn Museum, gift of Samuel P. Avery

26

27

28

a sometimes purplish lapis-lazuli blue, a dull red, yellow and white, to a dark green that on occasion approaches black. Even in the earliest *cloisonné* enamels there are examples of two colors being used in the same cell. Sometimes these blend with each other; at other times they remain distinctly separate. This is not considered a use of true mixed colors; but as techniques were refined, fragments of different colors eventually were mixed together. Green and yellow occasionally were combined. The most famous of the mixed colors, dating from later in the fifteenth century, was Ming pink, which is made from red and white fragments. For almost two centuries this was the only pink available to enamelers.

A progressively wider range and subtlety of shades evolved, however, and by the seventeenth century a dazzling palette of mixed colors existed, with as many as four colors combined in a single cell.

Early in the eighteenth century, rose pink enamel, derived from gold, was developed, and a beautiful style of floral decoration known as *famille rose* evolved around it. The sweetness of this rose pink and the vivid turquoise blue found in *cloisonné* of all periods (colorplate 10) are perhaps the most distinctive achievements in the color spectrum of Chinese enamelers.

In tracing the development of Chinese enameling, a progressive deterioration in quality can be observed. The design of the fifteenth- and much sixteenth-century work has a certainty of touch, with simple strong shapes in elegant patterns and clear singing colors. After the end of the Ming dynasty in the mid-seventeenth century, technical refinements gradually made mass production easier, and toward the end of the eighteenth century vast quantities of *cloisonné* were being produced by the Imperial workshops. There is a greater variety of designs, the enamels are glossier and smoother and achieve higher standards of technical excellence, but some of the early vigor is absent. Nonetheless, the models of birds and animals and creatures of fantasy found in the work of this period often have great charm. Finally, in the nineteenth century, design and decoration become fussy and the sure touch of the early craftsmen is gone.

In the eighteenth century, however, another very important development occurred, that of Chinese painted enamels, a last burst of glory before the art's decline. It is an interesting example of a collaborative effort between two sharply different cultures, for Chinese painted enamels did not evolve alongside the native tradition of *cloisonné* and *champlevé*. Rather, the technique was taught by visiting Europeans.

The emperor K'ang-hsi (reigned 1662–1722) was an enthusiastic patron of the arts. In about 1680 he established Imperial Palace workshops in Peking, which included *cloisonné* enameling among their techniques. By 1687 the technique of painting on enamels, which had

28.
One of a pair of Buddhist lion-dogs—the largest recorded pieces of Chinese *cloisonné* enamel—predominantly turquoise in coloring and measuring almost 2.70 m. (8¾ ft.) in height. Even though each piece was made in five parts, the individual elements are massive and would have required immense furnaces for firing the enamel. The high cost of producing such objects indicates that they possibly were made for an emperor to present to a famous temple. Late seventeenth or early eighteenth century. University of Pennsylvania Museum, Philadelphia

29.
A *cloisonné* dove-shaped wine vessel on wheels, with a trumpet-mouthed vase on its back. Modeled after bronze vessels called *chiu ch'e tsun* of the Han dynasty (206 B.C.–A.D. 220), it was intended to be passed around during the wine ritual. Close of K'ang-hsi period (1662–1722). Height: 55.8 cm. (21¾ in.). Brooklyn Museum, gift of Samuel P. Avery

Colorplate 11.
A large painted enamel dish with a scene taken from a European print. Painted enamels were introduced into China in the late seventeenth century by Jesuit missionaries, and their Western influence endured. Chinese, Ch'ien-lung period (1736–96). Diameter: 36.8 cm. (14¼ in.). Victoria and Albert Museum, London

30

been perfected in seventeenth-century France, was being brought to Peking by Jesuit missionaries.

The emperor's great interest in painted enamels was encouraged, for the Jesuits saw the advantages that might accrue, and a series of French painters and designers, some in holy orders, traveled to China to initiate Chinese craftsmen into the mysteries of painting on enamel. Some of the earliest designs reflect the conflicting tug of the two cultures: European shepherdesses in Chinese landscapes, cherubs interspersed with Buddhist emblems, Western buildings set among Chinese cliffs and waterfalls. But these incongruities add an endearing charm to the early works (colorplate 11).

Exquisite pieces were produced with *famille rose* decoration, while the *famille verte* palette, incorporating a wonderfully rich emerald green, is seen on a very few early pieces. In Peking and in Canton workshops, too, all types of objects were decorated with painted enamels, termed *yang-t'zu* (foreign porcelain). By the reign of Ch'ienlung (1736–96) prodigious quantities were being manufactured for export (plate 30). Objects in the pure Chinese taste, made for discriminating native connoisseurs, are rarer; usually they are heavier in make and finer in decoration (plate 31).

The early and mid-eighteenth-century work attained a delicate perfection, but later in that century the designs of the so-called Canton enamels became restless, fussy and overemphatic—familiar signs of an art in decline.

Throughout the nineteenth century vast quantities of enamels were made in Peking, Tientsin and Canton, mainly for export. During the twentieth century, with barely a pause for the Revolution, Chinese enamelers have continued to produce what have become mechanical and derivative enamels, apparently unconcerned at the paradox involved in seeking to copy again and again the aesthetic achievements of a despised Imperial past.

Japan   The Japanese refer to enameling as *shippo*, which means "seven treasures." The final effect of a piece of enameling was akin to seven precious things: gold, silver, emerald, coral, agate, crystal and pearl. Exactly when the Japanese began their seven-treasures work is once again a matter for scholarly debate. A small hexagonal enameled bronze plaque, thought to date from the Asuka period (A.D. 552–710), was excavated from a tomb in Nara prefecture, but most authorities agree that the earliest identifiable Japanese *cloisonné* was produced in Kyoto during the first decade of the seventeenth century, when the establishment of military rule put an end to the wars that had plagued Japan for so long. This was the beginning of the

30.
A cup and cover painted with examples of the "Hundred Antiques"—archaic bronzes and scholarly objects that were much admired by antiquarian Chinese scholars of the eighteenth century. Gilt-metal mounts. Chinese, Ch'ien-lung period (1736–96). Height: 15.2 cm. (5⅞ in.). Cooper-Hewitt Museum, in memory of Howard J. Sachs

31.
A hatstand painted *en famille rose* with motifs representing fortune and longevity. This is from the world's finest collection of Chinese treasures, which became one of the most dramatic adventures in the history of art. For sixteen years from 1933 a group of thirty scholars traversed 2,500 miles of wartorn central China with 20,000 cases filled with priceless objects from the Chinese Imperial collection. In 1948, at the direction of Chiang Kai-shek, the cream of the collection—in all 4,200 cases containing 60,000 pieces, among them 1,871 enamels—was taken to Taiwan. Red and gold lacquer stand. Ch'ien-lung period (1736–96). Overall height: 28.4 cm. (11⅛ in.). National Palace Museum, Taiwan

32.
A *tsuba*, or sword guard, in *cloisonné* enamel on bronze with a design of butterflies and crickets among grass. The majority of early Japanese enameling is on sword fittings; larger enameled objects, such as ornaments and vases, were not made in Japan until the nineteenth century. Maker: Kunikiyo, a swordsmith of the Echizen Seki School. Japanese, seventeenth century. Diameter: 7 cm. (2¾ in.). Brooklyn Museum

33.
A huge O-Daiko barrel drum is used offstage in Kabuki theatrical performances to create mainly naturalistic sounds, evoking whatever atmosphere the play demands. Kabuki was the popular theater of the Edo period (1600–1868) and the musical instruments used were those of the Noh drama. This *cloisonné* enameled bronze drum was never intended to be played; it was sent by Japan to the 1873 Vienna Exposition as a pacific gesture. In Japanese folklore, a fowl perched on a drum symbolizes peace. Height: 1.59 m. (5 ft. 2⅜ in.). Metropolitan Museum of Art, New York

highly organized and prosperous Edo period, which lasted until 1868. The country was ruled by Tokugawa shoguns, who presided from Edo (present-day Tokyo).

During this entire period the country was virtually isolated from the rest of the world, although some Dutch settlers were allowed to remain, operating their East India Company through small trading posts, and a certain amount of trade and artistic interchange was maintained with Korea and China. Fine craftsmanship flourished, but until the second half of the nineteenth century Japan remained sealed off from Western ideas and influences.

The art of *cloisonné* enameling may have been brought to Japan from China or Korea by Hirata Dōnin (1591–1646), who was the sole enameler employed in Kyoto by Shogun Tokugawa Ieyusa, and who was succeeded in this post by nine of his descendants in turn, until the late nineteenth century. At the Katsura Palace in Kyoto, enameled metal fittings attributed to Hirata decorate the sliding doors and lintels.

The Asaka City Museum has a crucifix decorated with *cloisonné* enamel that was once thought to date from the Momoyama period (1574–1600). This early origin is now in doubt as there is a possibility that it was made in the nineteenth century; it is in any case improbable that, if genuine, this would have been made later than 1614, when the anti-Christian edict was proclaimed. The famous silver and *cloisonné* mirror (discussed on page 39) in the Shoso-in at Nara is considered to be from the seventeenth century or later.

The earliest Japanese *cloisonné* may well have been done on sword mounts (plate 32), which is not surprising considering the military nature of the society. There is a group of cast-bronze sword fittings with *champlevé* enamel decoration in dull white, yellow and dark green that is thought to have been influenced by Korean work. It is termed Hirado after the Japanese west coast port that traded with Korea. Unrefined in appearance, Hirado enamels have been attributed to the seventeenth century, but they are more likely to have been made from the mid-eighteenth century onward. We know from the *Soken Kisho*, a book about swords published in 1781, that the Hirata family was active in the production of enameled sword fittings, but few examples of such work survive. Specimens of Japanese enameling before the mid-eighteenth century are exceedingly rare, and the best and most attractive examples date from the nineteenth century.

In 1839, the artist Kaji Tsunekichi (1803–1883) revived the art of *cloisonné* in Nagoya. For the rest of the century enameling came into its own as the Japanese evolved highly refined techniques such as "rainbowing" colors, dispensing with large numbers of cloisons and substituting large fields of enamel.

Japanese enamelers produced two types of work. From the 1860s, for the home market, small quantities of enamels were made based

33

on a subtle blending of subdued tones. Since enamels were considered to be precious objects, before the 1870s their sale had been barred to foreigners, but from the 1870s enamels also were produced for export in bright and brilliant colors, intended to appeal to what the Japanese thought of as a somewhat barbarian "foreign taste."

All manner of enameled objects have been made in Japan (plate 33) and designs vary a good deal, from abstract and geometric patterns to figural scenes containing animals and human figures. As in Chinese enameling, there is an enthusiastic use of naturalistic landscape and symbolic animals, birds and flowers. The dragon has magical attributes; the ho-ho bird symbolizes the majesty of the emperor, and the crane, longevity; the Kiri crest, based on the paulownia tree, is the badge of the military ruler or shogun; and the chrysanthemum, the national flower of Japan, is incorporated into the design of the Imperial crest.

Early Japanese work makes use of a variety of colors, but they are not so bright as those employed by the Chinese nor indeed as those in later Japanese enamels. Following the renaissance of Japanese enameling in 1839, only opaque enamels were used, in beautiful shades of red, green, lilac, pink, blue and yellow. From about 1870, opalescent, translucent and transparent enamels were increasingly employed, and this led to the distinctive Japanese discovery of "gold dust." Copper shavings or tiny fragments of silver foil were embedded in transparent enamel to send out gleams and glitters as they caught the light (plate 34). A few of the leading Japanese artist-enamelers had individual marks, which generally were engraved or stamped into the silver or copper base of a piece.

Toward the end of the nineteenth century a new school of enameling led by Namikawa Yasuyuki (1845–1927) was established in Kyoto. Its work raised Japanese enameling to fresh heights, combining technical expertise with deft artistic imagination (colorplate 12). A German chemist, Gottfried Wagner (1831–1892), who went to Japan in 1868 to develop pottery glazes, formed an association with Namikawa Yasuyuki, and together they researched and developed new enamel colors.

Some Japanese techniques approach those of painted enamels, since the wire work forming the cells is removed altogether before firing, and during the manufacture of a piece the procedure could be repeated with successive firings. This exceptionally difficult technique—called *musen shippo* ("cloisonless enamels") by the Japanese—was a specialty of the firm of Namikawa Sosuke of Tokyo (1847–1910).

Translucent enamels on a silver base were made by Ando Jubei of Nagoya, brother of the founder of the firm of J. Ando and Co., established in 1880 and still extant. Their chief artist was Kawade Shibataro, the grandson of Kaji Tsunekichi, an innovator who, in 1910, developed a new form of *plique-à-jour* (*shotai shippo*), creating a wholly translucent body (plate 35); he also produced enamels that had been embossed or worked in *repoussé*. Another original technique created by Kawade Shibataro was *moriage*, which consists of small areas of raised enamel, simulating the blossoms and leaves of plants.

By 1900 there were approximately two hundred enamelers in Japan, but mass production of cheap commercial wares led within a few years to the collapse of the market: by 1910 the entire industry was in decline. Today there are only a few principal manufacturers who produce both fine-quality and commercial enamels.

34.
Against a rust-brown gold-dust ground, the cover of this jar is decorated with a butterfly and a band of flowers. On the base, which has a predominantly turquoise ground, flower motifs and stylized leaves are set within panels with alternating gold-dust and black opaque enamel backgrounds. Japanese, late nineteenth century. Height: 9.5 cm. (3¾ in.). Minneapolis Institute of Arts

Colorplate 12.
A group of *cloisonné* vases made with silver cloisons on gilt copper, with silver and gilt mounts made by Namikawa Yasuyuki, an artist who excelled at the finest, most intricate designs; in the early years of the twentieth century he was the last remaining court artist practicing the art of *shippo*, as enameling is called in Japan. Japanese, late nineteenth century. Height of central vase: 15.2 cm. (6 in.). Victoria and Albert Museum, London

35.
A curved bowl on a small cylindrical foot in allover translucent *plique-à-jour* enamel with roses and buds in shades of orange, rose pink, purple and blue, between green leaves on a pale turquoise ground. Japanese, early twentieth century. Diameter: 12.5 cm. (4⅞ in.). Cooper-Hewitt Museum, gift of Mrs. Robert Ridgway

35

6 Russia

The movement of ancient tribes swept new ideas into Russia, which was geographically so placed as to bask in the influence of many diverse cultures. The Sarmatians—Persian nomads from the region east of the Caspian Sea—spread throughout western Russia to Europe during the second century B.C., bringing with them an Asiatic tradition of polychromatic decoration. Archeological finds show that gold and silver objects were then being inlaid with enamel, and by the third and fourth centuries A.D. *cloisonné* on gold and *champlevé* on copper were also being produced in various parts of Russia.

From the sixth century on, Russian work shows signs of Byzantine influence. In the eleventh and twelfth centuries the goldsmiths at Kiev were working in *cloisonné*, often employing silver, a departure from general Byzantine practice.

Not until the late seventeenth century was contact with western Europe finally established by Peter the Great (1672–1725). Then, as Russian rulers became aware of their nation's position as an eastern outpost of European civilization, they eagerly invited foreign artists and craftsmen to their courts. The Russian desire for things European accounts for the many examples of Western enameling to be found there.

Despite competing influences, certain parts of Russia evolved highly original styles. Indeed, Georgia should be considered separately, because of its geographical position between the Black and Caspian seas, sheltered by the Caucasus, and because of its distinctive culture and language.

Georgia enjoyed a long and unique history of metalwork: as early as the fifth century B.C. the Georgians had evolved a flourishing tradition of *repoussé* work, which combined well with enameling. Finds

Colorplate 13.
A Usolsk enamel bowl—displaying a type of enameling that is uniquely Russian. The piece is made of silver-gilt, with ten inset embossed lobes framed in fine twisted wire. The interior is painted with formal flowers and birds and, in the center, a river scene with two swans. On the exterior of the lobes shaped filigree enamel motifs are applied, with five portraits of men alternating with stylized flowers and birds. Russian, seventeenth century. Diameter: 20.3 cm. (8 in.). British Museum, London

36

from the Armazi tombs show that by the second century A.D. *cloisonné* enameling was established in Georgian jewelry manufacture.

At the end of the sixth century *cloisonné* enamels with Christian themes began to appear in Georgia (they arrived in Byzantium at the same time). The nature of the objects also changed; the bridles, necklaces, cups and lamps of pagan culture gave way to icons, crosses, triptychs and chalices.

From the twelfth to the fifteenth century Georgian enamelers achieved a marvelous range of icons. Mongol invasions had a disastrous effect on artistic activity, however, and during the sixteenth century Georgia was locked in combat with Turkey and Persia. The great age of Georgian enameling was over, although the craft continued to be practiced there (plate 36).

In Russia during the fourteenth and fifteenth centuries enamel was rarely used except as *basse-taille* decoration on large areas of metal. But in the sixteenth and seventeenth centuries Russian enameling took a tremendous leap forward. Its character changed completely, becoming increasingly dominant in the decoration of stylish objects such as *kovshi* (ceremonial drinking vessels) and scent bottles. Workshops in the village of Veliky Ustiug, in northern Russia, and at the Kremlin in Moscow developed sophisticated forms of filigree and skan enamel (plate 37). But the greatest surge of activity took place in the north of the country.

Solvychegodsk, an important trade center some five hundred miles east of St. Petersburg, was presided over by the Stroganovs, a wealthy merchant family. The influence of the Stroganovs in Russia could be compared to that of the Medici dynasty in Florence; during the sixteenth and seventeenth centuries they were the greatest power in the land and their patronage of the arts was legendary.

The Stroganov family collected works of art, attracted foreign craftsmen to Russia, and at Solvychegodsk they established workshops of silversmiths, icon painters and enamelers. Here, as elsewhere in Russia, painted enamels on a white background began to appear. Known as Usolsk work, it was similar to the technique developed around 1630 by the Frenchman Jean Toutin. A Persian influence also is present in much of the decoration, however, and some of the forms, such as a succession of concave lobes in the shaping of a bowl, are reminiscent of the metalwork of ancient Iran. These silver or copper bowls were beautifully painted with tulip and fritillary, twisting sinuous foliage and figural subjects (colorplate 13).

Peter the Great was an influential force in the development of enamelwork in Russia. He first visited France and England in 1698 and met the Swedish painter in enamels Charles Boit (1663–1727). Boit painted the czar's portrait. Evidently Boit created in him an enthusiasm for this form of miniature painting, for Peter began to invite European artists to Russia. Soon several French miniaturists were at court; many indigenous artists also took up this difficult art form, reaching an astonishing degree of proficiency and in some instances excelling the achievements of European masters. They produced portraits of religious subjects, rich merchants and royalty (plate 38).

During the eighteenth century diverse types of enameled objects were made. But in court circles, that sophisticated toy of the European *beau monde*, the gold box, became the most desirable of these objects. The St. Petersburg workshops produced lavish gold boxes, enameled and bejeweled, which the empress Elizabeth (1709–1762) would present to friends, protégés and lovers.

Catherine the Great (1729–1796), who described herself as having not just a love of the arts but "a downright gluttony" for them, appointed many artists to work for her. They included the enamel painter Barnabé-Augustin Mailly (1732–1793) and Jean-Pierre Ador (d. 1784), a Geneva-born goldsmith who came from France to Russia and worked in St. Petersburg from 1770 to 1785 (plate 39).

Ador was one of the most renowned goldsmiths of his day. A resplendent example of his work is the diamond-set gold box (now in the Smithsonian Institution in Washington) celebrating the *coup d'état* that brought Catherine to the throne in 1762. Made after 1770, its enamel panels, painted by the German miniaturist Karl August Kaestner, show the events leading to Catherine's enthronement. Kaestner was not the only German to make his mark on Russian enameling. Johann Gottlieb Scharff (active 1767–1808) carried the art of enameled and jewel-encrusted gold boxes to realms of glorious technical virtuosity.

37

36.
Folding box case for a small book of gospels. The Crucifixion with the Virgin, St. John and a third figure are painted in translucent and opaque enamels on gold. An inscription in cursive Georgian says, MADE IN 1687 BY EVNIHIDZE GUIORGUI. Georgian, seventeenth century. Height: 10.6 cm. (4½ in.). Walters Art Gallery, Baltimore

37.
A coconut cup mounted in skan enamel on silver. This style of enamel decoration became fashionable during the reign of Ivan IV (1533–84), a dedicated collector and patron of enamelers. Russian, late sixteenth century. Height: 11.6 cm. (4½ in.). Walters Art Gallery, Baltimore

38.
A painted enamel miniature portrait plaque of Peter the Great, Catherine I and their children, by Grigorij Mussikijskij (d. 1737). The eighteenth-century Russian miniaturists made a specialty of group portraits. Dated 1720. Width: 10.8 cm. (4¼ in.). Walters Art Gallery, Baltimore

38

39   40

From 1761 to 1776 the Popov factory at Veliky Ustiug near Solvy-chegodsk produced domestic wares and ornaments of silver or copper completely covered with white or blue enamel with encrusted silver *paillon* decoration (plate 40), and occasionally delicately painted with one or two colors. This style of decoration, which originated in the 1720s in the Fromery workshops in Berlin, is also seen on rare Russian enamels from the 1740s and 1750s.

In the nineteenth century there was a revival of interest in past designs and the Pan-Slavic movement became influential, promoting a taste for the bright colors and primitive forms of ancient Russian art. The famous firms of Fedor Ruckert, Paul Ovchinnikov and Ivan Khlebnikov made many types of fine-quality domestic items, from tea caddies and vodka sets to spoons in silver, enameled in *plique-à-jour* and filigree in traditional Russian styles (plate 41).

Much of this work, however, seemed garish to sophisticated European eyes. Their acclaim was to be reserved for a new name to emerge from Russia: that of Peter Carl Fabergé (1846–1920). In the Paris Exposition of 1900 he showed his work for the first time and won instant international admiration, receiving the *Légion d'honneur* and the title *Maître*.

The Fabergé family are thought to have been Huguenots who left France after the Revocation of the Edict of Nantes in 1685. By 1842 Gustav Fabergé had established a goldsmith's business in St. Petersburg, and four years later Carl Fabergé was born. After completing his formal education, Carl traveled widely around Europe, visiting London, Paris and Florence, and spending some time apprenticed to a celebrated German goldsmith, Friedmann, in Frankfurt-am-Main. He returned to Russia in 1870 and was made manager of his father's firm. It was an auspicious start, for by 1884 Czar Alexander III had granted his royal warrant to the House of Fabergé, and Carl's success was assured.

For the next three decades the workshops created a dazzling range of *objets d'art*, the most famous and spectacular of which were the celebrated Easter eggs made every year for the royal family (color-plate 14). Fabergé's virtuoso craftsmen fashioned exquisite enameled flowers in rock crystal vases (plate 42), but some products were functional as well as decorative. He made photograph frames, bell-pushes and cases for cigarettes, sealing wax, and so on. Yet much of his work was above all an artistic tour de force, with no functional role whatever—it simply celebrated privilege and taste. No wonder that when the Russian Revolution erupted, Carl Fabergé had to flee to Switzerland, and his workshops were closed down by the Bolsheviks. He died in Lausanne in 1920.

39.
A heavy gold potpourri, possibly a gift from Catherine the Great to Count Gregory Gregorievich Orlov, who directed the *coup d'état* that brought Catherine to the throne in 1762. The lid is surmounted by a crowned escutcheon bearing the blue enamel and gold monogram of Gregory Gregorievich. Modeled by Jean-Pierre Ador (d. 1784) in the style of a French porcelain vase, the object has accents of dark blue *basse-taille* enamel and medallions of opaque enamel painted *en camaieu* (in monochrome) in sepia on a white ground. Russian, dated 1768. Height: 28.3 cm. (11⅛ in.). Walters Art Gallery, Baltimore

40.
A Russian covered cup with an ornamental design in encrusted silver *paillon* decoration on a white enamel background. The piece is almost certainly from the Popov factory at Veliky Ustiug. Russian, 1770. Height: 27.5 cm. (10¾ in.). Walters Art Gallery, Baltimore

41.
The *kovsh*—a ceremonial drinking vessel— is a traditional Russian form, often elaborately decorated, intended for display rather than use. In *plique-à-jour*, edged with twisted silver-gilt wires, this example is enameled in translucent blues, white and red in a leafy geometric design. Russian, nineteenth century. Length: 11 cm. (4¼ in.). Cooper-Hewitt Museum, gift of J. Lionberger Davis

41

Fabergé's creations certainly provide a graceful and sometimes frivolous foil to the early Russian work, with its muscular barbarism. That two such different styles could have developed in the same country demonstrates the peculiar position of Russia, poised between East and West both geographically and culturally.

Colorplate 14.
The Fabergé Lilies of the Valley Easter Egg, presented by Czar Nicholas II to his mother, the dowager empress Maria Feodorovna, in 1898. The gold egg is of pink *guilloché* enamel, with green enamel leaves and pearl flowers. A geared mechanism enables a group of three rose-diamond–framed miniatures to rise out of the egg: Nicholas II in military uniform, flanked by his two eldest daughters, Olga and Tatiana. Cabochon rubies complete the embellishment of this tour de force of the jeweler's art, which is surmounted by a jeweled imperial crown. Fabergé workmaster: Michael Perchin; miniaturist: Johannes Zehngraf; marks: *MP, 56 anchors*. Russian, 1898. Height when open: 20 cm. (7⅞ in.). The FORBES Magazine Collection, New York

42.
A Fabergé spray of cornflowers. The flowers are enameled in blue, with gold stamens, stems and leaves, the pistils tipped with rose diamonds, set into a rock crystal vase carved to simulate water. Russian, 1896–1906. Height: 14.6 cm. (5½ in.). Virginia Museum of Art, bequest of Lillian Thomas Pratt

7 Scandinavia

ew Scandinavian enamels survive from earliest times. Eighth-century Swedish objects such as horse trappings have been found with red and yellow enamel decoration, and enameled articles combined with *millefiori* glass have been discovered in both Sweden and Norway. But these might have been taken there either through trade or by the Vikings as booty from England or Ireland.

Sweden Richly enameled chalices and patens, some of them dating from the first part of the fourteenth century, are found in Swedish churches. Many of these have been attributed to Stockholm goldsmiths, but their styling indicates that they most probably were imported from Germany.

One of the earliest examples of Swedish enameling by a known maker is the crown of King Erik XIV (colorplate 15), which was made in Stockholm in 1561 by the Flemish goldsmith Cornelius ver Weiden, active in Sweden from 1551 to 1561. The orb of Erik XIV (plate 43), also made by Cornelius ver Weiden, is the only known orb to be decorated with a terrestrial globe. Both objects are part of the Royal Treasury in Stockholm, which includes a wealth of enameled regalia.

Painted enamels were introduced into Sweden by the Frenchman Pierre Signac (1623–1684), who is believed to have studied in Paris with the renowned goldsmith Jean Toutin. Signac spent his life in Sweden, working from 1647 to 1684 for three successive sovereigns: Queen Christina (plate 44), Charles X Gustavus and Charles XI. His works consist of portraits and allegories, and he had a number of pupils, the most accomplished being Charles Boit, who became famous as a court painter in England in the late 1690s.

43.
The gold orb of King Erik XIV of Sweden, who was crowned at Uppsala in 1561. Made in Stockholm by Cornelius ver Weiden, the black enamel was added in 1568 by Franz Beijer in Antwerp. This was severely damaged in 1675 at the coronation of Karl XI; for the coronation of Adolf Frederick in 1751 the existing translucent blue enameling was done by the goldsmith Frantz Bergs. Swedish, 1561. Height: 16.5 cm. (6½ in.). The Royal Treasury, Stockholm

Colorplate 15.
The crown of Erik XIV of Sweden is decorated with unpolished opaque and translucent enamel and is set with emeralds, rubies and diamonds. The catalogue of the Royal Treasury states: "This Crown has been subjected to extensive alterations and a great deal of rough treatment over the years." It was dropped at a wedding banquet in 1568 and fell off Karl XII's head after his coronation in 1697. Made in Stockholm in 1561 by the Flemish goldsmith Cornelius ver Weiden. Height: 23.7 cm. (9⅜ in.). The Royal Treasury, Stockholm

44

Signac's successor at court was Elias Brenner (1647–1717), and another enameler active at this time both in Sweden and at the French court was Carl Gustav Klingstedt (1657–1734), known as *le Raphael des tabatières*.

Fine enameled work, especially gold watches (plate 45) and snuffboxes, continued to be made in Sweden throughout the eighteenth century. The boxes often bear all the signs of conventional Louis XIV snuffboxes and are difficult to ascribe to Scandinavia. But a certain lack of refinement has been identified as a distinguishing factor.

45

44.
An enamel miniature of Christina, queen of Sweden (reigned 1632–54), by the French artist Pierre Signac (1623–1684), who was appointed enameler to the Swedish court on January 1, 1647, when Queen Christina was twenty-one. Made in 1647. Height: 5.5 cm. (2⅛ in.). Nationalmuseum, Stockholm

45.
Two Scandinavian enameled gold watches that echo the techniques and designs of French and Swiss models. *Left:* by E. Lindgren, Stockholm (active c. 1740–50), a verge watch, its chased case enameled with colorful flowers and fruit. *Right:* by Isaac, Larpent & Jurgensen, Copenhagen, c. 1780, a *verge fusée* watch, painted in neoclassical style with a priestess sacrificing a bird, enameled predominantly in pink and blue shades. Eighteenth century. Fitzwilliam Museum, Cambridge

Denmark   There is a small group of *champlevé* enamels, eleventh-century in their iconography, which in recent years have been attributed to Denmark, although there appears to have been no continuing tradition from that period. Much debated by scholars, in the past these enamels were called Rhenish, but they are unlike all known Rhenish craftsmanship. A process of elimination—they bear no real relation to French, English or Irish work—has led to a Danish attribution as elements of their design are similar to those found in certain Danish objects and architecture. Their subject matter is biblical and their coloring brilliant; only nine objects thus attributed are known (eight boxes and one plaque), all now in famous collections.

In other respects, Denmark's history of enameling appears to echo Sweden's: sixteenth-century records suggest that enamel was used principally to decorate royal regalia. The richly enameled crown of King Christian IV, now preserved at Rosenbörg Castle, Copenhagen, was made by the Danish goldsmith Dirk Fyring in 1596; and the later regalia, dating from the 1660s, is also beautifully enameled.

In the seventeenth and eighteenth centuries Danish enamelwork was stylistically closely influenced by European trends, as was the Swedish. Paul Prieur (c. 1620–1683), who learned his craft at Geneva, was the son of a Parisian goldsmith. His signed miniatures date from 1645 until 1682, when he was working in London. Many of his subjects were royal, including the Danish kings Frederik III and Christian IV, King Michael Wisniowiecki of Poland and Charles II of England.

Many Continental artists were successful at the Danish court. But Joseph Brecheisen (1720–1770), a Viennese who had been working in Berlin when he was summoned to Copenhagen in 1757, achieved greater renown than his predecessors. He executed various types of enamels, including miniatures and gold boxes. In Copenhagen, from the late 1750s, enamel-on-copper snuffboxes became fashionable, and Brecheisen's work appears on many existing examples (plate 46).

At this time other enamel-on-copper boxes thought to have been made in Denmark were decorated with sketchily drawn playing cards, maps or stanzas from popular French songs—a fashion also adopted by German and English enamelers. But it would appear that the only native Danes who practiced enameling were Jörgen Gylding (1725–1765) and Hans Jacob Schrader (1718–1780) from Copenhagen.

After the Empire period, during which most of Europe reflected the all-conquering French style, the popularity of enameling declined

46.
A pair of enamel-on-copper boxes, painted inside the lids with miniatures of the sons of Denmark's King Frederik V: Prince Christian, aged eleven, who in 1766 succeeded his father to become King Christian VII; and Prince Frederik (in the inscription, Fredric), aged seven. The portraits are by Joseph Brecheisen, and the exteriors of the boxes are decorated in *trompe l'oeil* to simulate envelopes addressed *"Au Roy."* Danish, 1760. Width: 7.6 cm. (3 in.). Rosenbörg Castle, Copenhagen

in Scandinavia as elsewhere. But later in the nineteenth century, at the great international exhibitions held in London, Vienna, Paris and Budapest, highly accomplished Continental enamelers showed work inspired by Renaissance techniques, and this created a new enthusiasm for the art. In Scandinavia, an unexpected exuberance and vigor were manifest, above all in Norway, which until then had had little part to play in the history of the art.

47.
A glass claret jug, the silver filigree mounts decorated with multicolored *plique-à-jour* enamel. Norwegian, made in 1894 by J. Tostrup of Oslo. Height: 19 cm. (7⅜ in.). Kunstgewerbe Museum, Berlin

Norway   During the eighteenth and nineteenth centuries in Norway enameling played a minor role in the training of jewelers: making an enameled ring was one of the tests for an apprentice to become a master. From the Empire period, small-scale work remains, such as jewelry and frames for miniatures. But the important period of Norwegian enameling began around 1880.

Although its character originally was Russian in inspiration and some craftsmen may have traveled there from Russia, the industry was essentially unique to Norway. The tremendous development in the nineteenth century was the result of national expansion and pride in a sense of prosperity, and Norway achieved worldwide acclaim for its enameled works of art. It was the era of Ibsen and Grieg, a flowering of the creative arts, the moment at which Norwegian culture burst into life.

During the 1880s the firm of J. Tostrup in Oslo began large-scale production, first of *champlevé* enamels, then of *cloisonné* and filigree, and from the 1890s, of *plique-à-jour* (plate 47). Initially the designs were based on past Russian peasant work, but gradually essentially Norwegian styles evolved. At J. Tostrup, Torolf Prytz (1858–1938) was mainly responsible for the artistic side of production. A leading technician was Emil Saether, who produced *plique-à-jour* bowls, lamps and vases of Art Nouveau design.

A magnificent goblet made by Saether was shown in Paris in 1900, where it received a *Grand Prix* (see colorplate 26). When Saether died later that year, Tostrup stopped producing ornate works of this nature but elaborate spoons and other small objects in highly colored *plique-à-jour* were made in his atelier. Other manufacturers mass-produced enameled souvenirs, for which a simplified form of *plique-à-jour* was devised.

In 1914 a centenary exhibition was held in Oslo to commemorate the establishment of Norway's free constitution. There Jacob Prytz, Torolf's son, showed a collection of enamels that involved a new technique, employing brilliant transparent colors on engraved and chased backgrounds. Today Jacob Prytz's daughter, Grete, is a leading Norwegian enameler.

The second distinguished Norwegian firm of enamelers and silver-smiths is David-Andersen, which in 1887–8 installed modern equipment for a wide variety of techniques (plate 48). An important David-Andersen silver-gilt casket, decorated with *champlevé* and *plique-à-jour* enamel (plate 49), was presented by Norwegian admirers to Louis Pasteur on his seventieth birthday in 1892. One of David-Andersen's leading designers, working for the firm until 1910, was the Bohemian-born Gustav Gaudernack. Both Tostrup and David-Andersen are among the rare firms that, aside from wartime interruptions, have succeeded in maintaining constant production for over a century.

48.
David-Andersen of Oslo produced elaborate enameled tableware with great panache. This silver-gilt and *cloisonné* enamel coffeepot incorporates dragon's-head motifs, reminiscent of Viking emblems. Norwegian, 1891–3. Height: 16.5 cm. (6½ in.). Kunstindustrimuseet, Oslo

49.
A silver-gilt casket made by David-Andersen in Oslo to be presented to Louis Pasteur on his seventieth birthday. It is a masterpiece of enameling, with *champlevé* sides and *plique-à-jour* on the lid. Norwegian, 1892. Length: 25 cm. (9¾ in.). Musée Pasteur, Paris

49

8 The Continent of Europe

The Holy Roman Empire gave a coherence to Europe that for many years was reflected not only in the homogeneity of its political and religious organization but also in its art. Princes intermarried, priests and scholars traveled, pilgrimages and Crusades were undertaken and trade thrived. Dynamic interaction such as this helped to spread ideas associated with the gradually unfolding international styles: Romanesque, Gothic, Renaissance, baroque, rococo.

The Byzantine influence was strong—at its height in the tenth and eleventh centuries it was copied in many European countries. As early as the seventh century, Byzantine enameling (begun in the previous century) probably was imitated in northern Italy by Lombard craftsmen, and in England the ninth-century Alfred jewel appears to have been influenced by Byzantine *cloisonné* work.

The life of Empress Theophano (c. 955–991) exemplifies how the Byzantine web of intermarriage affected the spread of artistic influence. A Byzantine princess who married King Otto II, she brought enamelers with her to Germany. At Byzantium magnificent gold *cloisonné* work was produced, and from the Ottonian period (936–1002) the technique flourished in Rhineland Germany at Essen, at Mainz, and particularly at Trier in the workshops of Archbishop Egbert (937–993). It was also practiced successfully in the Mosan area of Lorraine in eastern France.

During the Carolingian and Ottonian empires, which spanned the mid-eighth to the early eleventh centuries, Byzantine-inspired enamelwork was combined with carved ivories and elaborately mounted precious stones.

Another noblewoman, Countess Gertrudis I from Brunswick, founded the great Guelph Treasure, from which nine rare articles are now preserved in the Cleveland Museum of Art (plates 50 and

Colorplate 16.
A detail from the right wing of the triptych house-altar of Duke Albrecht of Bavaria. Assembled from many superbly composed parts, the ebony altar is mounted with figures, ornaments and plaques of enameled gold, the three-dimensional elements decorated *en ronde bosse* and the flat areas with translucent *champlevé*. Precious jewels complete the enrichment. Attributed to the goldsmith Abraham Lotter, the Elder of Augsburg. German, 1573–4. Height of wing illustrated: 20.3 cm. (8 in.). Residenz der Schatzkammer, Munich

50

50.
The Gertrudis portable altar, from the Guelph Treasure, is unique in being the only known portable altar in gold. On a core of oak, the lid is inset with porphyry, and on the sides arches of *cloisonné* enamel frame embossed figures of saints. Semi-precious and glass-paste stones are set into the borders. German, Lower Saxony, c. 1040. Length: 26.7 cm. (10⅜ in.). Cleveland Museum of Art, purchase, The John Huntington Art and Polytechnic Trust

51.
A silver-gilt arm reliquary on a kernel of oak, with multicolored *champlevé* enamel borders depicting Christ and the twelve apostles. Attributed to a follower of Eilbertus of Cologne, a master goldsmith and enameler who exerted a profound influence on the early twelfth-century school of Lower Saxony. German, Hildesheim, c. 1175. Height: 51 cm. (19⅝ in.). Cleveland Museum of Art

51). She presented precious objects in memory of her husband to the cathedral of St. Blasius at Brunswick. When the countess died in 1077, many other exquisite enamels were added to the collection. Later, in the early twelfth century, relics from the Second Crusade were acquired, and even until the end of the fifteenth century further additions were being made to the Guelph Treasure. In this way individuals and families have contributed to Europe's dazzling cultural heritage.

During the early twelfth century there was a distinct movement toward *champlevé* enameling on copper or bronze, and a consequent neglect of *cloisonné* (although it was not entirely abandoned, appearing in the finer details of some designs, when it often was used together with *champlevé*). Gold was mainly suitable for enameling small articles and jewelry; but cheaper, base metals enabled much larger areas to be decorated.

It is not clear whether the revival of *champlevé* started in Spain, on the Rhine or the Meuse, or in Limoges, but by the mid-twelfth century craftsmen in all these centers were producing enamels in the Romanesque style. There was a distinctive school of enameling in Spain around that time, showing some signs of Muslim influence and principally organized around the town and abbey of Silos.

Rhenish and Mosan monasteries and workshops produced magnificent, creative works that were characterized by their bold designs and brilliant coloring. Few names are known of enamelers from these regions, but among the most eminent were Godefroid de Claire of Huy (plate 52), who in the 1150s purportedly made a reliquary triptych for the monastery of Stavelot, near Liège; Eilbertus of Cologne, who influenced the enameling at the famous goldsmiths'

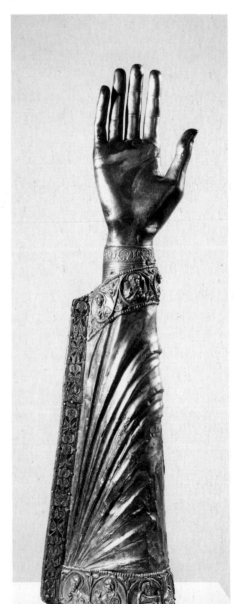

51

52

52.
Two plaques, depicting the Baptism and the Crucifixion, in which the *cloisonné* and the *champlevé* techniques are combined, possibly by Godefroid de Claire, a leading early medieval goldsmith. Plaques such as these were used to decorate altars. French, Mosan, c. 1150–75. Width: 10.2 cm. (4 in.). Metropolitan Museum of Art, New York, gift of J. Pierpont Morgan, 1917

53.
Nicholas of Verdun, an outstanding twelfth-century goldsmith, is thought to have designed the huge Shrine of the Three Holy Kings in 1181. The largest shrine to have survived from the Middle Ages, it was not completed until c. 1220–30. The gold figures at the front depict Christ in Majesty and, below, the Virgin and Child and the Three Magi, after whom the shrine is named. German, Rhenish, twelfth to thirteenth century. Length: 210 cm. (81⅞ in.). Cologne Cathedral

53

54.
A spectacular example of a three-dimensional cast figure combined with *champlevé* enameling, this portrayal of St. James is from an altar frontal in the abbey of Grandmont in the Haute-Savoie. School of Limoges, thirteenth century. Height: 29 cm. (12 in.). Metropolitan Museum of Art, New York, gift of J. Pierpont Morgan, 1917

Colorplate 17.
A reliquary *chasse*, or chest, in *champlevé* enamel on gilded copper, from a church in Billom, Puy-de-Dôme, thirty-five miles east of Limoges. Decorated with scenes from the life of Christ; the heads of the figures are in relief, the bodies engraved against a blue enameled background. French, Limoges, thirteenth century. Height: 38.7 cm. (15 in.). Walters Art Gallery, Baltimore

55.
A thirteenth-century Eucharistic dove that might have been suspended above an altar, possibly as an allusion to the Holy Spirit. *Champlevé* enameling on copper. French, Limoges, thirteenth century. Length: 26 cm. (10⅛ in.). Musée de Cluny, Paris

school at Hildesheim in Saxony; and Nicholas of Verdun (plate 53), who produced the superb triptych altar at Klosterneuberg near Vienna in 1181.

Champlevé enameling reached its apogee in Limoges at the turn of the twelfth to thirteenth century. At the beginning of the twelfth century enameling was practiced at various French monastic centers, such as Conques in the southwest, Grandmont in the Haute-Savoie (plate 54) and Saint-Martial at Limoges in western France. Then, gradually, lay establishments were founded in and around Limoges, thereby centralizing production.

By the second half of the thirteenth century the Limoges workshops were highly commercial, manufacturing vast quantities of *champlevé* enameled goods for civil and religious purposes (plate 55). These Limousin products were exported throughout Europe and beyond until the late fourteenth century. It was a rare church treasure that did not include an ecclesiastical object enameled in Limoges—hence the wealth of specimens that have come down to us (plate 56).

The outstanding characteristics of the Limoges School at this time were deep, glowing, opaque enamel colors applied to richly gilded metal (colorplate 17). On many Limoges enamels gilt foundations were engraved in the technique known as *à fond vermiculé*: stylized branches and decorative motifs surrounding enameled panels, flowers, stars and rosettes.

Limoges *champlevé* enamels can be roughly divided into two groups: in the first, earlier group, enameled figures were set against a gilt background; in the second, conversely, gilt figures were reserved in an enameled (usually blue) field, and lines within the gilt figures—features, limbs, folds of draperies—were incised and filled with enamel, somewhat in the manner of niello. A striking effect was also achieved by the application of gilt-cast three-dimensional heads or whole figures (see plate 54). Further embellishments took the form of filigree ornament, embossed motifs and precious stones.

Enamel plaques were fixed to perishable books, caskets and so on; frequently the objects themselves have decayed and disappeared, but the enamel triumphantly survives. Some twelfth- and thirteenth-century Rhenish, Mosan and Limoges enamels rival the excellence of the best Byzantine work. Toward the end of the thirteenth century, however, there are signs of hasty and cheap manufacture. The art slowly declined, and the events of the Hundred Years' War, which in 1370 led to the sacking of Limoges by the Black Prince, brought about its final collapse.

In the late thirteenth century the new enameling technique of *basse-taille* was developed in Italy. The earliest surviving example—the chalice of Pope Nicholas IV—was made in Italy by Guccio della Mannaia around 1290. It is now in the treasury of the church of St. Francis at Assisi.

55

Basse-taille was practiced in many centers including Paris (plate
58), Valencia in Spain, Constance and Basel on the Upper Rhine
and Vienna. An outstanding example is the Royal Gold Cup of
the Kings of France and England (colorplate 18). But the most
glorious work came from Italy, especially Florence and Siena, where
the form originated. By the early sixteenth century the great period
of *basse-taille* enameling had passed, although its popularity endured
in southern Germany, principally in Augsburg, and in Spain until
the mid-seventeenth century.

During the late fourteenth and early fifteenth centuries European
goldsmiths were inspired to create newer, more daring forms of
enameling. At this time, *plique-à-jour*—perhaps the most mysterious
effect of all—was devised. Few early specimens remain of this highly
skilled technique, which creates a delicate framework of silver or gold
wires, similar to a cobweb, its filaments upholding fragile slivers of
brilliant, vitreous colors (plate 57).

A decade or so earlier another most significant development was
the innovation of painted enamels, first produced in Flanders and
soon taken up in Venice and France. A small medallion, probably
made in the Netherlands about 1425, is considered to be the earliest
surviving specimen of a painted enamel (plate 59).

By 1470 painted enamels on copper were being made in Limoges,
and by the early sixteenth century this famous center was once again
in the forefront of enameling, enjoying its second great period.
Nardon (Léonard) Pénicaud (c. 1470–c. 1542) was one of the earliest masters, and several members of his family followed. Others were
the anonymous enameler or enamelers known by the unexplained

57.
The Merode beaker is a rare example of
early fifteenth-century *plique-à-jour* enameling. The bands of enamel on the silver-gilt
beaker and cover are between plain areas
that are delicately pounced with birds, fruit
and flowers. The edges are finished with
twisted wire and punched decoration. Of
Flemish or Burgundian origin, c. 1430.
Height: 17.5 cm. (6⅞ in.). Victoria and
Albert Museum, London

58.
A silver-gilt reliquary shrine made for
Queen Elizabeth of Hungary (1300–1381),
consort of King Charles Robert from 1320
to 1342, and daughter of King Vladislav
of Poland. The tiny statuettes of the twelve
apostles, saints and musical angels are
enameled in *basse-taille*: the faces are in
flesh tones and the robes amber, topaz and
emerald green, with touches of opaque
vermilion, against a cobalt blue background.
The wings of the altarpiece open to reveal
a three-dimensional group, comprising the
Holy Virgin, the Child Jesus and two
virgins. Presumed to have been made in
Paris, about 1340–50. Height: 25.4 cm. (9⅞
in.). Metropolitan Museum of Art, New
York, Cloisters Collection

58

59.
Two views of a medallion that probably is the earliest surviving example of a painted enamel. The subject, known since 1323 as the Ara Coeli vision, depicts the Virgin and Child as an apparition seen by the emperor Augustus. It is painted in gold, white and dark blue enamel on a silver base. Netherlandish, c. 1425. Diameter: 5.1 cm. (2 in.). Walters Art Gallery, Baltimore

60.
Pierre Reymond was the leading maker of decorative tableware in sixteenth-century Limoges. The subject matter of this ewer, Melchisedech and Abraham, is from Genesis 14:18–19. Painted *en grisaille* on a black ground with faint flesh tones and gilded details. French, Limoges, c. 1555–60. Height: 32 cm. (12½ in.). Walters Art Gallery, Baltimore

59   60

name of Monvaërni, to whom some fifty works are attributed; the Limosins—Léonard, who excelled at portraiture (see colorplate 2), in 1548 was given the title of King's Painter by Henry II of France; Pierre Courteys; Jean de Court; Susanne de Court, the only recorded female enameler at Limoges, and Pierre Reymond (plate 60).

Contrary to the earlier, *champlevé* period, when the most vivid colors were used against shining gold foundations, the dominant shades of Limoges painted enamels were dark and somber. They reflected the pessimism prevalent at the time of the Reformation and also followed Spanish style, which at this time favored oppressive, rich decoration.

About 1535 *grisaille* (gray-toned) decoration was introduced. It maintained its popularity for several decades, until superseded by painting in vibrant colors, as seen in the works of Susanne de Court and the Limosin family. Paintings and prints by the great artists and engravers of the day were frequently copied or adapted by enamelers

Colorplate 19.
An unidentified enameler, known as the Aeneid Master, produced a series of plaques for which many designs were woodcut illustrations from an edition of Virgil's *Aeneid*, published in Strasbourg in 1502. There were 114 plates in the book and 74 extant enamel plaques depict subjects from them. This one shows *Aeneas Founding the Town of Acesta in Sicily* and is copied from Plate 61. French, Limoges, c. 1530? Height: 25.2 cm. (10 in.). The James A. de Rothschild Collection at Waddesdon Manor, Aylesbury, Buckinghamshire

61.
An engraving from Sebastian Brandt's edition of Virgil's *Aeneid*, published in Strasbourg in 1502, the subject of the Limoges enamel plaque above. British Museum, London

62 and 63.
A Venetian ewer and dish, painted in deep blue, green and white enamel with allover gilt enrichments. By the thirteenth century enameled glass was being made in Venice, and for centuries the city was the principal source of supply for enamel colors to craftsmen from many centers. It is therefore surprising—considering the artistry of the Venetians—that their output of enamel objects was neither significant nor sustained. Italian, Venice, early sixteenth century. Height: 29.5 cm. (11½ in.). Wallace Collection, London

(colorplate 19 and plate 61), and many specially commissioned pieces bore family crests.

Plaques were the first objects to be made at Limoges. Most often the decoration of these was religious in nature: they could be assembled in sets in metal frames to form polyptychs—small multiple-paneled altar screens. But from the 1530s all types of luxurious domestic and ecclesiastical objects were produced. Painted enameling was also practiced from the fifteenth century at Aragon and Catalonia in Spain, at Venice (plates 62 and 63)—long famous for the glassmaking at Murano—and indeed throughout northern Italy; but nowhere was the standard of Limoges work equaled.

In France toward the end of the fourteenth century an incomparable new technique originated: encrusted enameling, known as *émail en ronde bosse*. Three-dimensional gold figures or embossed reliefs, usually small, although a few were larger, were decorated with enamels (colorplate 16). The taste for encrusted enamel spread to all parts of Europe, and it continued as a genre through many changes of style—Gothic to Renaissance to baroque.

When a technique as splendid as this became fashionable through the exchange of gifts between church and court leaders, it was adopted by goldsmiths and jewelers in many countries (plate 64), but without

62

63

64.
A gold pendant with a pelican enameled *en ronde bosse* and set with pearls, rubies and emeralds. The subsidiary pendant baroque pearls were added in the seventeenth century. German, c. 1600. Height: 8.9 cm. (3½ in.). Cleveland Museum of Art, gift of Mr. and Mrs. Severance A. Millikin

65.
A dragon-shaped ewer and cover in carved rock crystal with enameled gold mounts. Each rock crystal element is attached to the body with a gold ring *champlevé* enameled in black with foliate scrolls. Italian, Saracchi workshop, Milan. Height: 24.8 cm. (9⅝ in.). Toledo Museum of Art, Toledo, Ohio, gift of Florence Scott Libbey

Colorplate 20.
The magnificent enameled gold reliquary bust of Ladislas IV, king of Hungary (reigned 1272–90). Gold was mined in large quantities in Hungary from the thirteenth to the fifteenth century (more than in any other European country) and important gold objects such as this were produced. The blue filigree enameling is superb. Each motif, framed by twisted wire, has tiny gold stars surrounding formal quatrefoil patterns. Hungarian, c. 1405. Height: 64.7 cm. (25¼ in.). Gyor Cathedral, near Budapest

marks or specific provenance, the attribution of examples is largely speculative. *Émail en ronde bosse* was produced throughout Europe, wherever fine enamels were made.

Italian goldsmiths excelled at this and other enameling techniques. Among the names that became famous were the Saracchi family of five Milanese brothers (plate 65), who were followed by three more from the second generation; Ambrogio Foppa, called Caradosso (c. 1446–1526), who was in the papal service, and, preeminently, Benvenuto Cellini. Cellini left us a clear account of his life, ideas and methods of manufacture, including enameling, in two books—the *Autobiography* and the *Treatises on Goldsmithing and Sculpture*. He created many masterpieces of the goldsmith's art, although the sole surviving gold piece known to have been worked by him is an enamel-decorated saltcellar made in 1543 for Francis I of France.

Perhaps the most sumptuous Renaissance objects of all were the tazzas, vases, caskets, ewers and all manner of decorative pieces that were carved in precious materials: lapis lazuli, red jasper, porphyry or rock crystal. After being carved and polished—treatments requiring skills of the highest order—these were engraved and fitted with gold or silver enameled mounts.

Paris, Florence, Milan and Nuremberg were celebrated centers for cutting and polishing hardstone, as well as enameling. A famous Italian family, the Miseronis, working at the Hapsburg imperial court

Colorplate 21.
Examples of a rare enameling technique, *en résille sur verre*. The German artist-engraver Valentin Sezenius is thought to have been the foremost practitioner of this technique. The medallion on the left, the Nativity, is after his own engraving of the subject; the pendant, right, depicting angels telling the shepherds of the Nativity (the scroll inscribed: "Gloria in Excelsis Deo") is also attributed to Sezenius; its frame is nineteenth century. Shown actual size. Paris or Prague, late sixteenth to early seventeenth century. Wallace Collection, London

66.
The front of a painted enamel-on-gold locket depicting a naval engagement with "H. Toutin" inscribed above, and on the reverse an inscription that, translated, states "Henri Toutin, master goldsmith at Paris, made it, 1636." The inside cover and the back (neither shown) respectively portray *Diana and Actaeon* and a military siege. Height: 4.5 cm. (1¾ in.). British Museum, London

workshop in Prague, excelled at the genre. Goldsmiths in many European cities were adept at mounting, but the finest mounts were made in Antwerp.

In the early fifteenth century Venice is thought to have reintroduced filigree enamel, which had been employed in earliest times. This technique spread not only to the rest of Italy but also to eastern Europe, where it was cultivated most successfully in Hungary (colorplate 20). Although many enameling methods had been used in Hungary, filigree enameling was adopted with such enthusiasm that it achieved a national identity, becoming essentially Hungarian, and as such it was imitated elsewhere.

By the first quarter of the sixteenth century the great period of Hungarian filigree had ended, but the technique continued to be practiced until the seventeenth century and was revived in the nineteenth century in Russia and Norway.

Another distinctive Hungarian form is known as Transylvanian enamel, although it was practiced throughout Hungary. Most often used for domestic articles and jewelry, this technique is akin to *cloisonné* but differs in that the silver base plate can be shaped as a flower, leaf, scroll or similar motif, with the edges turned up and indented to simulate twisted wire, forming a frame to contain the enamel. Transylvanian enameling flourished until the beginning of the eighteenth century.

In Paris, and possibly in Prague, the rare technique known as *en résille sur verre* was developed. Until recently this method of enameling was believed to have been practiced for approximately forty years only (from about 1600 to 1640), but recent research has led to the conclusion that it was in use as early as 1570–80. The name of Valentin Sezenius, a German artist-engraver, is particularly associated with this method (colorplate 21). Several of the rare examples that exist match engravings attributed to him, a few of the latter being dated from 1619 to 1625.

Toward the end of the sixteenth century Limoges painted enamels had suffered a catastrophic decline: there were reports of important objects being sold for derisory sums though production continues there to this day. But elsewhere in France the art was to evolve in a more sophisticated form.

From about 1630, vigorous new schools of enamel painting developed in Blois, near Orléans, in Paris and also in Geneva, following an innovation attributed to Jean Toutin. At these places, where there was a thriving tradition of watchmaking, enamelers began to work in gold instead of copper, laying white enamel onto the gold, to be finely painted with enamel colors.

Miniature figural subjects, landscapes and flowers now decorated every visible surface of the watchcases and dials. This was a time when the exquisite small watch was esteemed a most fashionable gift,

67

68

to be exchanged by royalty and exalted dignitaries, as well as by less august people.

Jean Toutin died in 1644, and no work unequivocally attributable to him has been recorded. But his two sons, Jean II and Henri, were both fine enamelers. Henri (plate 66) established himself in Paris and in 1636 painted and signed a portrait of Charles I of England, the earliest surviving example of the genre.

Miniature portraits in enamels were also the specialty of the greatest of the Geneva enamelers, Jean Petitot (1607–1691). He and Paul Prieur were both citizens of Geneva who became famous in northern Europe. Petitot traveled to England, possibly in 1636, and worked for Charles I; he returned to France around 1644, and by 1650 was established in Paris. Prieur settled in Denmark, but is known to have worked in London, since he signed and dated one miniature: "Prieur à Londres, 1682."

Two other eminent enamelers of Geneva were Pierre Huaud (1612–1680) and Jean André (1646–1714), both of whom founded families who contributed significantly to the practice of the art. Huaud had three sons, and all three were appointed enamel painters to the Great Elector of Brandenburg, who became King Frederick I of Prussia in 1701 (plate 67). Jacques Bordier (1616–1684) and Jean-Étienne Liotard (1702–1789) were two more famous Genevese enamelers who traveled throughout Europe practicing their art.

From the seventeenth century onward, all over Europe marvelous painted enamel objects were made in copper and in gold (plate 68). In some cases, it might seem, the more trifling the object, the greater the joy taken by the enameler in transforming it into an exhilarating *jeu d'esprit* (plate 69).

67.

A cup and saucer, enameled on copper and painted with genre and mythological scenes including *Cupid and Psyche*, *Venus and Adonis* and *Mars and Venus*. Signed: "Les Frères Haud [sic] Pin." Swiss, early eighteenth century. Diameter of saucer: 11.4 cm. (4½ in.). Victoria and Albert Museum, London

68.

A gold cup and cover, the body painted with scenes of the naval battle between the Dutch and the English off Chatham, Kent, in 1667. The Dutch government ordered three identical cups to be presented to the three commanders of the Dutch fleet. This one belonged to Admiral Adriaansz de Ruiter. Made by Nicolas Loockemans of The Hague (active 1648–70). Height: 30 cm. (11¾ in.). Rijksmuseum, Amsterdam

69.

In seventeenth-century Holland small domestic objects and exquisite personal accessories were decorated with enamel:

Top left: Gourd-shaped gold scent bottle, painted with multicolored flowers and set with a band of rubies. Height: 5.1 cm. (2 in.). Rijksmuseum, Amsterdam

Center left: Gold case for scissors, painted with multicolored flowers. Height: 8.3 cm. (3¼ in.). Victoria and Albert Museum, London

Right: Knife and fork handles painted in dark green, royal blue and red on a white background. Length: 10.8 cm. (4¼ in.). Fitzwilliam Museum, Cambridge

Bottom: Enchanting tiny gold pomander, shown closed and open, painted in white, violet, blue and green and set with diamonds, emeralds and rubies. Height: 4.2 cm. (1⅛ in.). Rijksmuseum, Amsterdam

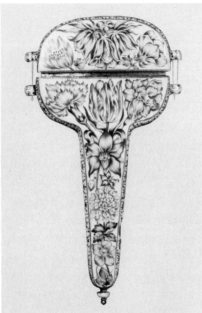

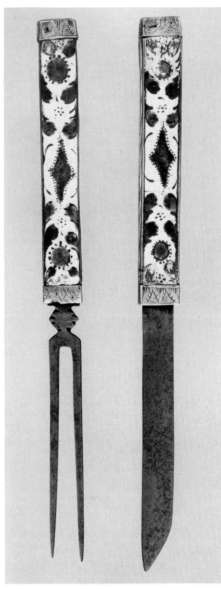

69

70

71

70.
Around the second decade of the eighteenth century Alexander Fromery in Berlin devised a unique style of enameling that also was practiced in France and later was copied in Russia. It consisted of encrusted gold *paillon* decoration on a white enamel background: this snuffbox lid is a typical example. German, c. 1745. Length: 8.3 cm. (3¼ in.). Victoria and Albert Museum, London

71.
A medallion painted in monochrome blue with the inscription, translated, "Cobalt Blue from Aragon, Barcelona, 1783." This was possibly made as a sample of a factory's production of a specific oxide for coloring. The steel frame is set with faceted blue glass stones. Spanish, eighteenth century. Width: 10.9 cm. (4¼ in.). Victoria and Albert Museum, London

The Germans produced some of the most glorious examples of the genre. The main centers of German enameling in the first half of the eighteenth century were Augsburg and Berlin. The Fromery family, who were gunsmiths and goldsmiths working in Berlin, had originally been French Huguenots, like so many other émigré craftsmen. Alexander Fromery, who took over the firm on the death in 1695 of his father, Pierre, devised the characteristic Fromery style. This consisted of shaped, embossed pieces of gold foil applied to a field of white enamel, at times this being the sole decoration, as in the case of many small boxes (plate 70); at other times it was combined with scenes or motifs painted in colors on the white enamel ground. The gold encrustations, called *paillons*, often were impressions taken from medals; some of those used by Fromery were rubbed off medals made by Raymond Falz, a Swede who worked in Berlin and Augsburg. Recent research at the Victoria and Albert Museum, London, has revealed that many enamels of this type, hitherto attributed to the Fromery factory, in fact were made in France, probably in or near Paris, where in 1777 there was a central guild of *paillon*-makers.

Not all of Fromery's enamels bore *paillon* decoration, even though this style is so closely associated with his name; many of them were exquisitely painted. One of the most skilled German enamel painters, Christian Friedrich Herold (1700–1779), worked with Fromery until 1725, when he moved to the Meissen porcelain factory, where he specialized in painting harbor scenes. He became one of the most important enamelers of his time, and probably continued to execute commissions for Fromery even after he left his employ. At this time it was customary for porcelain factories and enamelers to send out *blanks*, to be painted by artists, known as *Hausmaler*, in their own homes. This practice accounts for the fact that the same artist's hand can be discerned on both porcelain and enamel from the period. The Fromery workshops were active until about 1780, and Fromery-type enamels were widely appreciated and also imitated, particularly in Russia (see plate 40).

During the second half of the eighteenth century cheaper enamel-on-copper boxes and plaques were made throughout Europe (plate 71), although German workshops were the most prolific. Vienna was a flourishing center of enameling. Two brothers, Christoph and Johann von Junger, were active there from about 1764 to 1790. They produced, among other things, enamel-on-copper domestic tableware that imitated the fashionable Sèvres porcelain.

In eighteenth-century Europe snuff-taking was more than just a passing fad. Snuffboxes were clues to social standing, the most prized boxes being made of gold. The French court, which had nourished the arts ever since Francis I had patronized Cellini and Leonardo da Vinci, reached new heights of extravagance under Louis XV, and boxes of great beauty and fantastic technical virtuosity were created.

Gold boxes were made in all shapes and sizes for rouge, sweetmeats and patches (black "beauty spots" made of taffeta or paper) as well as for snuff, and in a glorious variety of enameled designs—portraits, landscapes, birds, beasts, plants and formal decoration. *Chinoiserie* was popular, originally influenced by the designs of Jean and Claude Bérain, then, in a lighter vein, by Watteau, Audran and Pillement.

Over 240 Parisian goldsmiths are known to have made gold boxes. Among the most prolific of those whose elegant *oeuvres* were enameled were Jean Ducrollay and Jean-Joseph Barrière (colorplate 22), Noël Hardivilliers, Jean Frémin, Jean George, J. E. Blerzy, Charles le Bastier and J. M. Tiron. Many virtuoso painters of enameled gold boxes were active at this time in Paris, and one of the most accomplished was Charles-Jacques de Mailly (1740–1817).

After the storming of the Bastille, the age of the gold box was apparently over. Boxes now were made in wood, ivory, papier-mâché or base metal. Around 1804, however, the taste for luxurious objects revived, albeit in moderation. Gold boxes were made in the Empire style with portraits painted in enamel, some of them of Napoleon, which the emperor presented as gifts.

The German gold boxes produced at Dresden and Berlin were more often of hardstone than enamel. Frederick the Great (1712–1786), who lived at Sans-Souci in Potsdam, was an inveterate snuff taker and collector of boxes. It was under his patronage that the manufacture in Berlin of snuffboxes enameled *en plein* in the Parisian manner was initiated. A distinguished painter in enamels who worked there was Frederick's principal designer and engraver, Polish-born Daniel Nicholas Chodowiecki (1726–1801).

Colorplate 22.
Eighteenth-century enameled gold objects of vertu, in the Wallace Collection, London.
   Top left: A Viennese oval snuffbox with domestic interior scenes by Philipp Ernst Schindler (1723–1793).
   Top right: A German cartouche-shaped snuffbox with a diamond-encrusted thumbpiece, with allegorical scenes by John William George Kruger (active in Berlin 1755–68).
   Bottom, clockwise from center: A scuttle-shaped snuffbox with bouquets of summer flowers (Paris, 1756–7); a rectangular snuffbox, made in Paris, 1749–50, by Jean-François Breton (active 1750–91); a navette-shaped shuttle, made in Paris, 1772–3, by Jean-Joseph Barrière (active 1750–93), with genre scenes and panels of *faux* malachite; a shell-shaped snuffbox with peacock-feather motifs, made in Paris, 1743–4, by Jean Ducrollay (1709–after 1761).

72.
A nineteenth-century reproduction of a sixteenth-century Limoges enamel-on-copper dish, painted *en grisaille* with *The Rape of Hippodamie* and signed L. Dalpayrat. Louis Dalpayrat (1844–1910) was an accomplished French enameler who went to London to lecture on enameling. French, dated 1883. Diameter: 21.6 cm. (8⅜ in.). Victoria and Albert Museum, London

In eighteenth-century Vienna, Philipp Ernst Schindler (1723–1793; see colorplate 22, top left) was one of the most admired enamelers of snuffboxes. Enameled snuffboxes were also made in Russia, Sweden and Denmark, but few seem to show any distinctive national characteristic. English boxes often had a style of their own, yet the influence of France was always predominant.

In Switzerland, however, there were exceptional developments toward the end of the century. Until about 1775–80 the Swiss confined their enameling principally to watches. But by 1790 workshops in Geneva were exporting boxes decorated with enameled paintings. The application of a final, thick layer of transparent enamel protected the paintings, which became ambitious and large-scale as a result.

Favorite subjects were lakeside scenes and, in the popular neo-classical style, heroic tableaux copied from painters such as Jacques-Louis David. The most elaborate boxes and watches were made for Oriental markets; they were formed in the shape of animals, butterflies (colorplate 23) or fruit, and painted in vivid colors. In Geneva in 1789, 77 enamel painters were recorded; by the beginning of the nineteenth century there were 215. Such was the progress of the industry. It is remarkable that few of these artists signed their work. We do know, however, of Jean-Louis Richter (1766–1841) and the Roux brothers, all of whom were landscape painters.

After the second quarter of the nineteenth century most enameling seems to have looked back to the past, either remote or recent, for its inspiration. At this time Vienna was more important as an enameling center than any other European city except Paris or Geneva. This was partly because the formation of the Austro-Hungarian Empire in 1866 gave Vienna access to Bohemian and Transylvanian silver and gold, but it was also partly attributable to plentiful cheap labor.

A demand for expensive-looking objects in Renaissance style developed, and the firm of Hermann Ratzersdorfer obliged, leading a Renaissance revival and producing works of art for the vitrines of a new breed of collectors. At its height, Viennese enameling was indeed prolific. Although it catered to the fashionable trade of visitors to the spa at Carlsbad, increasingly the subject matter was ill-chosen for the piece it was to decorate, resulting in a pastiche of past designs. World War I finally put an end to the production of these lavish and often ostentatious objects.

In France a medieval revival took place in the early to mid-nineteenth century, inspired by the novels of Victor Hugo, which created a vogue for the Gothic taste. But the Renaissance proved a more fruitful source of inspiration—and one more generally popular with the public.

Around 1840 yet another revival began, of Limoges painted enamels. And by the 1850s the Sèvres porcelain factory, with state subsidy, was manufacturing replicas of sixteenth-century wares. A decade later

73.
A replica of the enameled gold reliquary altar cross of the Hungarian king Ludwig I, the original of which was made shortly after 1370, when Ludwig was elected king of Poland. This was stolen from the Geistliche Schatzkammer (Spiritual Treasury) of the Austrian Empire between 1860 and 1875 and was copied by Salomon Weininger, a Viennese antique dealer and master faker. The original was replaced in the Schatzkammer about 1930. Height: 43 cm. (16¾ in.). Hungarian National Museum, Budapest

several Parisian ateliers were in similar production and this work flourished into the 1880s. At this time Louis Dalpayrat (1844–1910) was one of the leading enamelers making Limoges reproductions (plate 72), which had achieved great success in England. As a result of their popularity Dalpayrat went to London in 1885 to lecture on enameling at the South Kensington Museum (now the Victoria and Albert). Before the end of the century, however, the vogue for these heavy, somber objects was spent.

European enamelers and jewelers were obsessed with Renaissance designs. Some nineteenth-century craftsmen carried their fascination with the past a little too far. It was acceptable and even admirable for the renowned Italian Castellani family to produce jewels in imitation of ancient designs; quite another matter for Salomon Weininger (1822–1879) to fake rare museum acquisitions.

Weininger worked in Vienna, where he was entrusted with the restoration of various precious objects, the majority of them enameled, in the Geistliche Schatzkammer—Spiritual Treasury—of the Austrian Empire. In fact, Weininger made copies of the pieces, returned the copies to the treasury, and sold the originals (plate 73). For many years his copies were accepted as genuine. The full extent of Weininger's activities only became apparent as the originals turned up in salerooms after the deaths of the collectors who had bought them from him. Although he died in the Austrian state prison, Weininger's works are now exhibited as remarkable artifacts in their own right.

If forgery and ancestor worship were aspects of the nineteenth century and its enamelers, mass production and the mixed blessings of technology were the other side of the coin. The firm of Brüder, Gottlieb & Brauchbar, located in Brunn in Austria (now Brno in Czechoslovakia) and active from the late 1890s until 1914, produced crudely decorated enameled steel beakers and mugs to celebrate great state occasions.

During the nineteenth century French manufacturers turned out vast quantities of enameled buttons, as well as trinket boxes in imitation of Sèvres porcelain. Samson et Cie of Paris marketed a wide range of replicas of eighteenth-century works, from English enamels to Chinese Ch'ien-lung objects. Not until the Arts and Crafts movement in Britain and the development of Art Nouveau in France, Vienna and Norway would enamelers go beyond mere imitation of previous forms to find true inspiration in the past.

9 Great Britain

From at least the first century A.D., British Celtic enamelers used the *champlevé* technique on flat or rounded surfaces, sometimes in several colors. A primitive form of *cloisonné* was also used to fill studs on bronze objects with red enamel, or *cuprite glass* (see plate 11). Celtic enameled horse trappings (see colorplate 5), jewelry and vessels such as skillets and vases are among objects dating from the first to the third century A.D. found in the British Isles.

Evidence of enameling in Britain is lacking from the early third until the fifth century, when enamels were imported into England from Ireland. The Irish produced bronze bowls and jewelry decorated with *champlevé* enamel. *Millefiori* glass also was incorporated into their work (plate 74)—at Garranes and Lagore *millefiori* sticks were found in the remains of workshops dating from the fifth to eighth century.

The Sutton Hoo Ship Burial, excavated near Woodbridge, Suffolk, in 1938–9, included bronze hanging bowls decorated with *champlevé* enamel combined with *millefiori* glass, thought to date from A.D. 625–30 (see colorplate 3). A rare and most important early example of *cloisonné* is the late ninth-century Alfred jewel, in which the enamel rests under a layer of rock crystal within an ornate, filigree gold setting bearing the inscription AELFRED MEC HEHT GEWYRCAN (Alfred had me made). It is certain that the jewel was made for King Alfred (849–899), probably as part of an aestel (a pointer) or a crown. But there is controversy as to whether the piece is of English or foreign origin.

With the spread of Christianity, pagan burials ceased—unfortunately, from the archeologist's point of view, as graves were no longer supplied with jewelry and useful objects for the afterlife. Since *cloisonné* was produced mainly on gold, very little of that has survived

Colorplate 24.
The gold Phoenix jewel of Queen Elizabeth I of England. It has been suggested that the Virgin Queen adopted the legendary phoenix as an emblem of herself. The similarity between the gold bust and a miniature of Elizabeth by Nicholas Hilliard, dated 1572, has caused this jewel to be dated 1570–80. The reverse *(left)* shows a phoenix in flames under the royal monogram, and the whole is framed by an enameled garland of red and white Tudor roses and green leaves. English. Diameter: 4.6 cm. (1¾ in.). British Museum, London

74

75

either. A few *champlevé* brooches in copper and bronze have come down to us, however, to suggest that enameling was a continuing tradition in England during the ninth and tenth centuries. The Norman Conquest in 1066 brought a distinctive Continental influence into all aspects of the arts in Britain.

A limited number of *champlevé* objects have been tentatively identified as twelfth-century English. Among the most notable are the Whithorn crozier of around 1200 (plate 75) and the Masters plaque, in the Victoria and Albert Museum, which dates from about 1150–60.

As soon as enamels began to be made at Limoges, they were exported to England. By the early thirteenth century English bishops were specifying that the ciboria that held the Host were to be made in silver, ivory or *opere Limovitico* (Limoges enamelwork). The number of Limoges *champlevé* objects found in England and the references to them indicate that their importation most probably would have discouraged the progress of any native school of English enamelers. The increasing international uniformity of late Gothic style in ecclesiastical arts, however, makes it difficult at times to assign thirteenth- and fourteenth-century objects to particular schools.

It follows that most English work about which there can be little doubt or debate tends to be secular. Some is heraldic, such as the enameled stall plates of the Knights of the Garter in St. George's Chapel, Windsor, produced between 1368 and 1499. Tombs were often covered with enameled shields, and *champlevé* continued to be used for horse trappings, perfume cases, saltcellars, ewers, cups, horn mounts, other domestic articles and jewelry.

There are many documentary references to enameled objects owned by the English Crown, or presented by the king to foreign princes and dignitaries, but few examples survive. Those that do suggest that the use of translucent *basse-taille* enamel was popular during the fourteenth century.

In the fifteenth century there was a decline in the arts, the usual inevitable accompaniment to wars and rumors of wars—1453 saw the final English defeat at the end of the Hundred Years' War, and in 1455 the Wars of the Roses began. Even so, enameling may have continued in England during this period. The encrusted or *en ronde bosse* technique was developed in France toward the end of the fourteenth century; there was a very close connection between the courts of the two countries, and it is most probable that this technique, which was extremely popular with goldsmiths and jewelers, was also practiced in England (plate 76).

During the Tudor dynasty (1485–1603), however, enameling flourished. One of the jewels given by Henry VIII (1491–1547) to Catherine Howard was described as having "beades of goldesmytheswercke of another sorte ennamuled with white/havying a pillor of golde ennamuled blewe . . ." according to an inventory of 1587.

74.
A bronze brooch with three areas of red *champlevé* enamel inlaid with platelets of multicolored *millefiori* glass. This was found in 1933 during an excavation at the Ballinderry Crannog (a crannog is an ancient dwelling on a man-made island in a lake). Irish, seventh century. Length of pin: 18.3 cm. (7⅛ in.). National Museum of Ireland, Dublin

75.
A crozier depicting iconographical scenes in red, blue, green, yellow and white *champlevé* enamel on copper, discovered during excavation in 1957 at Whithorn Priory in Galloway, southeast Scotland, on the site of a medieval cathedral. The four largest figures represent bishops: the one with a halo is presumed to be St. Ninian, founder of the church at Whithorn. Sixteen other figures holding books or scrolls possibly portray the four major and twelve minor prophets. English or French, c. 1200. Height: 19 cm. (7⅜ in.). National Museum of Antiquities of Scotland, Edinburgh

76.
The Swan jewel was found on the site of a Dominican priory at Dunstable, Bedfordshire, in 1965. It is a superb example of opaque white enamel used *en ronde bosse* over gold. The swan motif was adopted by many aristocratic families, and this jewel probably belonged to a person of noble birth. English or French, fifteenth century. Height: 3.2 cm. (1¼ in.). British Museum, London

76

Henry's vain daughter Elizabeth I (1533–1603) was lavished with presents of jewelry by her admirers, who sometimes had incorporated into the design a subtle clue as to their identity. In the case of the Earl of Leicester, it was his symbol of a ragged staff with "blacke men clyminge upon ragged staves," and in the case of her French suitor, Alençon, a little enameled gold frog. Elizabethan enamels abound in this kind of symbolism.

Enameled pendants and lockets, such as the unique Phoenix jewel (colorplate 24) were made bearing Elizabeth's portrait. Nicholas Hilliard (1547–1619), the distinguished miniaturist, painted portraits of the queen and other notables that frequently were enclosed in elaborate cases, sometimes densely enameled in a mixture of *cloisonné* and *champlevé* in a wide range of colors and often richly bejeweled as well. These might be sent as gifts among the nobility and occasionally served to inform suitors of the features of their intended brides.

In the second half of the seventeenth century there was a new departure in English enameling, the cast-brass work known as Surrey, or Stuart, enamels, thought to have been made at Esher. Fewer than one hundred examples exist, among them many different articles ranging from fire-dogs and candlesticks (plate 77) to bit-bosses and badges. The object was cast in brass with a low-relief surface pattern, often engraved to improve the sharpness of the design, and then the recesses were filled with enamel. The use of color is fairly restrained, black and white or blue and white being the most common combinations, with green also appearing and occasionally red or yellow.

There is an exotic feeling to some of the designs on these pieces, which has been attributed to the influence of Daniel Diametrius, who

77.
A Surrey enamel candlestick, known as the Warwick candle, with vine-tendril motifs typical of Charles I decoration, and with birds, dogs and hares enameled in turquoise blue and white. English, c. 1650–60. Height: 25.5 cm. (10 in.). British Museum, London

founded the brass mill at Esher in 1649. He and his partner, Jacob Momma, were Germans, but Diametrius may have been Greek in origin. There is also a similarity to Russian, Bavarian and other Continental work of this type, as well as a marked Italianate note in much of the detail.

During the seventeenth century events were taking place on the Continent that were soon to have an even more decisive influence on English enamels. Religious persecution forced many craftsmen to flee from France, and England provided a relatively liberal haven. In about 1636, the Frenchman Jean Petitot arrived from Paris, bringing with him a technique he had learned from Jean Toutin that was to revolutionize English work: the painting of miniature portraits in enamel. This technique, evolved by Toutin around 1630, used gold as a base for painting in delicate colors on a white enamel ground. Petitot's output was prolific (in the collection at Windsor Castle alone there are almost three hundred of his miniatures), although in some cases he is reputed to have painted only the faces—the clothes, hands and so on being "studio work." Other skilled enamelers followed, several of whom achieved fame owing to their work for royalty and the aristocracy.

The Swedish-born Charles Boit, who arrived in 1687, painted Peter the Great's portrait during the czar's first visit to London in 1698. Boit's pupil, Christian Friedrich Zincke (1687–1770), who arrived from Dresden in 1706, was patronized by George II. Many distinguished English miniaturists also chose the medium of enamel, among them Samuel Cotes, Richard Crosse, William Prewitt, Gervase Spencer and Henry Bone. These highly accomplished artists were to have a lasting effect on the decoration of English objects of vertu. Their style of meticulous, detailed enamel painting was the perfect medium not only for portraiture but also for the decoration of gold watches, snuffboxes and other small, precious articles. The designs of such luxuries, however, often were inspired by (or copied from) famous French paintings by great masters, such as Watteau, Fragonard and Boucher (see colorplate 1).

The neoclassical style was also high fashion at this time, and crumbling temples and draped Grecian ladies found their way into many an English landscape. Swiss-born George Michael Moser (1706–1783), a founder of the Royal Academy and its first Keeper from 1768 until his death, was a famous goldsmith and enameler represented by signed pieces in both genres who painted in the Grecian style; among others were James Morriset, Gabriel Wirgman and William Charron.

But another method of making charming enameled trinkets for the *haute monde* was about to develop. From the mid-1720s in England, there was widespread production of painted enamel-on-copper dials for watches and clocks. At first these were simply decorated with Roman or Arabic numerals; later landscapes and figural subjects were added.

English manufacturers were curiously slow to make the imaginative leap from watch faces to other small objects such as snuffboxes and patch boxes. Not until the arrival around 1740 of elegant copper enamel boxes from Berlin and an irresistibly chic variety of small enameled trifles from Paris (although these were on gold) were the first enamel-on-copper boxes attempted by English craftsmen.

Some critics cursorily dismissed enamelwork on copper as irredeemably inferior to that on gold and therefore barely warranting consideration. But the potential of these charming bibelots was swiftly realized, and by the late 1740s many manufacturers were beginning to make the new little boxes.

By 1760 the production of painted copper enamels was a highly successful, established industry. At various periods during the second half of the eighteenth century enamels were made at Battersea on the Thames in London; in other London workshops; at Bilston and Wednesbury in South Staffordshire; at Birmingham and at Liverpool (plate 78). Conditions were perfect for manufacturing to thrive: the Industrial Revolution facilitated mass production and was also

78.
English enamels of the 1770s attributed to different sources: a basket-shaped box, possibly from Birmingham, the lid painted with a harbor scene; an egg-shaped thimble case credited to Bilston, painted with classical landscapes; and a small oval Liverpool locket, transfer-printed in brown with Polly Peachum from *The Beggar's Opera*, written by John Gay in 1728 and revived in 1778, the year of publication for the engraving from which this transfer was taken. Height of locket: 3.3 cm. (1¼ in.). Bantock House Museum, Wolverhampton, England

79 and 80.
"Bottle tickets" or "wine labels," many of
them after engravings by Simon-François
Ravenet, were a specialty of the Battersea
factory (1753–56). Generally in the form of
wavy escutcheons, their subject matter
usually involved cherubs engaged in activities appropriate to the named wine or spirit.
English, Battersea, c. 1754. Width: 7 cm.
(2¾ in.). Victoria and Albert Museum,
London

creating a prosperous bourgeoisie eager to cultivate the *bon ton*
of their social superiors. The French court provided a scintillating
model, and the eighteenth-century taste for delicacy, frivolity, wit
and in later decades sentimentality admirably suited the creation of
these tiny *jeux d'esprit* with their miniature scenes, jokes and messages.

The factory at York House, Battersea, although operational for
barely three years, from 1753 to 1756, acquired a fame that still
endures after more than two centuries. It was founded by Stephen
Theodore Janssen (1705–1777), a colorful character who acquired
many civic distinctions (he was Lord Mayor of London in 1754–5),
but the demise of his enameling enterprise ended with his bankruptcy.
The business was probably established to develop the new technique
of transfer printing, an invention claimed by an Irishman, John
Brooks, who for a mere six months had been Janssen's partner at
York House. Being among other things Master of the Worshipful
Company of Stationers, Janssen was in a good position to acquire suitable paper for transferring the engravings.

Simon-François Ravenet (1748–c. 1814), a distinguished French
engraver who had been invited to England by Hogarth in 1744,
worked at Battersea on designs for snuffboxes, plaques and wine labels
(plates 79 and 80). He produced some beautiful scenes from classical
mythology as well as portraits of eighteenth-century royalty (plates
81 and 82) and of aristocratic and religious figures.

The exquisite elegance and brilliance of Ravenet's designs are the most remarkable achievement of the Battersea factory. No other transfer-printed decoration on enamel can equal his work. Battersea enamels usually were printed in monochrome in varying tones of red, purple, sepia or charcoal on a white background; only rarely were they overpainted in soft, translucent colors.

London ateliers are thought to have produced some exceptionally elegant painted enamels, including portraits of famous beauties (plates 83 and 84), *chinoiserie* scenes (plate 85) and even some risqué subjects—all suited to a sophisticated metropolitan taste. There was obviously an affinity between London enamelers and the famous Chelsea factory, just across the Thames from Battersea. The hand of the same flower painter is discernible on both enamel and porcelain objects; furthermore, Chelsea porcelain bonbonnières (containers for sweetmeats) were fitted with enamel lids, and enamel scent bottles had Chelsea porcelain stoppers.

One group of boxes ascribed to London is transfer-printed with calendars for the years 1757–60 (plate 86); some of these are signed: "Anth. Tregent, Denmark Street." It is possible, however, that Tregent was not the manufacturer but a shopkeeper who had enamels bearing his name made elsewhere—a common practice with all types of goods, since a London mark added considerable prestige to provincial merchandise. One of the fascinating aspects about so many enamels from all countries is the lack of marks or signatures. This is

81 and 82.
A pair of portrait medallions, transfer-printed in crimson, in gilt-metal frames. *Left:* William Augustus, duke of Cumberland (1721–65), third son of King George II. *Right:* King George II (reigned 1727–60). Taken from engravings by Simon-François Ravenet. English, Battersea, c. 1754. Height: 8 cm. (3⅛ in.). Victoria and Albert Museum, London, Schreiber Collection

83

84

85

83.
A snuffbox decorated with Frans van der Mijn's (1719–1783) unknown lady, taken from the mezzotint in plate 84. The enameled portrait is by an unidentified miniaturist of distinction whose work occasionally can be identified on other fine-quality enamels. London or Bilston, 1760–5. Length: 6.1 cm. (2⅜ in.). Bantock House Museum, Wolverhampton, England

84.
A mezzotint of an unknown lady by Richard Purcell (1736–1766), after a painting by van der Mijn, inscribed: "Painted for Robt. Sayer at the Golden Buck in Fleet Street." The painting on the lid of the box is taken directly from this picture. English, c. 1760. Bantock House Museum, Wolverhampton, England

85.
Two *chinoiserie* painted oval plaques, their subjects copied from *Livre de Chinois*, a book of engravings by P. C. Canot taken from designs by Jean Pillement, published in London in 1759. These plaques were bought in Paris from Mme Flaudin by Lady Charlotte Schreiber on November 17, 1881. Her journals reveal that she had to pay "a large price for them, 20 guineas." London or South Staffordshire, c. 1765. Width: 17.8 cm. (7 in.). Victoria and Albert Museum, London, Schreiber Collection

86.
A calendar snuffbox for the year 1758, the lid and base each transfer-printed in charcoal with six months of the year and the sides with feast days. Boxes similar to this occasionally are signed "Anth. Tregent, Denmark Street." Tregent, born in Geneva in 1721, was one of three brothers who settled in London. Birmingham or London. Length: 6.5 cm. (2½ in.). Bantock House Museum, Wolverhampton, England

87.
A ring-stand with an inset watch, the latter engraved "Hughes, London," possibly William Hughes of 119 High Holborn (active 1769–94). The enamel is painted with pastoral scenes and, on the reverse, a still life; the green enamel of the upper part is decorated with gilded rococo scrolls and flowers. South Staffordshire, c. 1770. Length: 12.5 cm. (4⅞ in.). Cooper-Hewitt Museum, bequest of Katherine Strong Welman

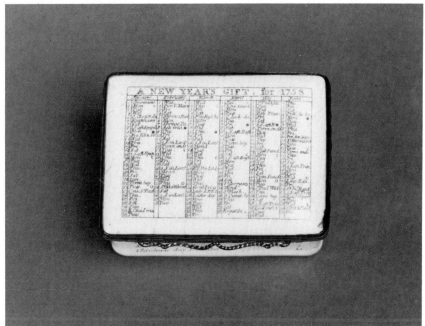

86

87

particularly perplexing in the eighteenth century, when it was customary for the marks on most manufactured articles to declare their origin. The attributions of various types of English enamels to the specific places or areas of manufacture described here are those that are generally accepted; absolute proof of origin of an English enamel object is exceedingly rare.

There was a tradition of metalwork at Bilston in South Staffordshire from the late seventeenth century. French refugee craftsmen settled there, and the enameling industry gradually developed. By the 1760s there were eighteen enameling workshops in Bilston. The enamels produced there and at nearby Wednesbury (both towns were part of the manufacturing area known as the Black Country) are considered to be among the most lavishly decorated domestic and personal objects made in eighteenth-century England (plate 87; see also colorplate 1): candlesticks, vases (plate 89), *étuis* (cases fitted with small personal accouterments; colorplate 25, bottom left, and plate 88) and sumptuous caskets containing tea caddies (plate 90) (tea being a precious commodity at that time, to be kept under lock and key).

But apart from such magnificent pieces, much simpler articles, instantly attributable to Bilston, are charming small snuffboxes or patch boxes with inscriptions (colorplate 25, top left), messages of love and courtship or friendly greetings. Scent bottles and bonbonnières were also made at Bilston in the form of three-dimensional animals (colorplate 25, top right), birds, flowers and fruit.

A characteristic of South Staffordshire is the delightful range of intense and delicate colors imitative of Sèvres porcelain, including the elusive rose pink made from gold oxide, known as *roze* in eighteenth-century and *rose Pompadour* in nineteenth-century France, and as *rose du Barry* in England. In 1780 a cheaper version of this pink was derived from chrome and tin, and became known as English pink by the Continental enamelers who imported it.

The subject matter of South Staffordshire enamel design was generally lighthearted: amusing scenes, flowers, birds and butterflies. Many of these subjects were copied from books of engravings, the most popular being *The Ladies Amusement: Or Whole Art of Japanning Made Easy*, published between 1758 and 1762 by Robert Sayer of London, which included over 1,600 different designs. The work could be hand-painted, transfer-printed in monochrome or transfer-printed and overpainted.

Birmingham was equally famous for its enamel production. The city was the most organized and prolific center for the manufacture of all types of metal goods: iron, steel, brass, silver, gold and, among countless commodities, copper forms for enameling. These "blanks," as well as the metal mounts and enamel colors (plate 91) produced in Birmingham, would doubtless have been supplied to enameling workshops elsewhere.

In 1751 John Brooks (later to become Janssen's partner at Battersea) made his first unsuccessful application to patent the technique of transfer printing from Birmingham. In the early stages of its development this method of decoration was used extensively there (plate 92). On the majority of Birmingham enamels, colorful decoration—

Colorplate 25.
English eighteenth-century enamels from Bantock House Museum, Wolverhampton, England.

Top left: Two transfer-printed and overpainted Bilston boxes, typical of the romantic trifles and souvenirs of visits that were exchanged as keepsakes in Georgian times. The lower box depicts the Dripping Well, a petrifying spring at Harrogate in Yorkshire.

Bottom left: A gilt-metal and enamel chatelaine—an appendage intended to be worn on a lady's belt, to hold keys and personal accouterments. Attached are two egg-shaped boxes for sweetmeats and an étui containing the implements shown below, in plate 88.

Top right: Three painted bonbonnières—a frog, a pug's head and a lion couchant.

Bottom right: Extraordinary mustard pots in the form of knights in armor were made during the 1770s and 1780s. A military trophy is painted within a reserve in the deep turquoise ground; on the back there are sprays of summer flowers.

88.
The contents of the pink étui illustrated opposite: an ivory slip, bodkin, ear spoon, nail file, penknife, pencil and scissors.

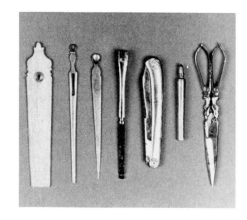

89.
An ornamental ovoid urn, painted in colors on a white background with *Rebecca at the Well* and on the reverse (not shown) *David and Abigail*, both after paintings by the Venetian artist Jacopo Amigoni (1675–1752). South Staffordshire or Birmingham, c. 1765. Height: 47.6 cm. (18⅝ in.). Metropolitan Museum of Art, New York, gift of Irwin Untermyer, 1964

classical landscapes, flowers or portraits—covered the entire area on a white background, without the frames, panels, cartouches and raised white or gold scrolling and enrichments that so often were present on South Staffordshire enamels.

At Liverpool, monochrome transfer printing was the specialty, particularly for portraits of famous people and characters from the theater (see plate 78). In July 1756 a young Liverpool engraver and

90

90.
An enamel-on-copper casket, lined with velvet and fitted with two tea caddies and a sugar canister. Every side of each object is painted with different pastoral scenes, some of which include Italianate classical ruins—the influence of the eighteenth-century Grand Tour of Europe. Gentle-folk and peasants are engaged in country pursuits in these idyllic settings. South Staffordshire or Birmingham, c. 1765. Length of casket: 21 cm. (8¼ in.). The James A. de Rothschild Collection at Waddesdon Manor, Aylesbury, Buckinghamshire

91.
Abraham Seeman was an enameler as well as a supplier to the trade of enamel colors. This advertisement appeared in *Aris's Birmingham Gazette* on September 16, 1751. Bantock House Museum, Wolverhampton, England

---

ABRAHAM SEEMAN,
ENAMELLING-PAINTER,
At Mrs. Weston's in Freeman-Street, Birmingham,

MAKES and Sells all Sorts of ENAMELLING Colours, especially the Rose Colours, and likewise all Sorts for China Painters, at reasonable Prices.

N. B. Most of the eminent Painters of Birmingham, Wednesbury and Bilson, have made use of the above Colours to their Satisfaction.

91

---

printer, John Sadler, with his partner Guy Green, announced that they were transfer printing ceramic tiles. As this was but two months after the sale following the failure of York House, Battersea, there is speculation that their equipment might have emanated from that source. Plaques and boxes marked "J. Sadler, Liverp: Enaml" are among the rare English enamels that bear a maker's name.

From the 1790s standards deteriorated as manufacturers exploited

the public craze for pretty enameled objects. Social changes had their effect, too: snuff-taking declined and with it the need for snuffboxes. Much of the appeal of enameled trinkets lay in their novelty, and novelty is not an enduring characteristic. By the 1830s the taste for this type of object had disappeared. The last recorded enameling workshop in the Midlands closed in 1840.

The Great Exhibition of 1851 at the Crystal Palace inspired British manufacturers to produce various gold-plated household articles decorated with enamel in a totally different manner, and the British firm of Elkington competed with its French counterpart, Barbedienne (see endpapers), in the production of elaborately enameled and gilded ornaments, many of them inspired by Moorish and Middle Eastern models.

The Italian Castellani and Giuliano families, both famous as jewelers, established businesses in London in 1860. Their designs and manufacturing methods, including enameling, were inspired by jewels of centuries past.

The English enthusiasm for all things enameled caused one French enameler, Louis Dalpayrat (see plate 72), to visit London in 1885. Among those attending his lectures at the South Kensington Museum was Alexander Fisher (1864–1936), the son of a potter from Staffordshire, who became England's premier enameler at the turn of the century (plate 93). His designs were influenced by the Renaissance, and during this period many artists were inspired to follow his style of enameling.

This conscious—some might say precious—cultivation of the past is in striking contrast to the way English enamels had been produced a hundred years earlier in the heyday of the eighteenth century. Then a bustling society had joyfully celebrated the tangible benefits of the newest manufacturing advances. It had celebrated them in an art form that embraced politics and mythology, nature and culture, the prince and the popinjay; and mass production had meant that an ever-widening public, made affluent by the fruits of the Industrial Revolution, could share in that enjoyment.

93

10 The Development of Art Nouveau and Enamels Today

Art Nouveau From the mid-1800s developments were taking place that would gradually pave the way for the heady excitement of Art Nouveau. First, a series of important exhibitions took place, at which designers hungrily devoured one another's ideas and discovered undreamed-of aesthetic inspiration from the art of ancient civilizations. The 1862 International Exhibition in London was vitally important. Here the Japanese showed their works of art for the first time to an astonished and ravished European public. Western designers reveled in the spareness of line and the deliberate asymmetry of Japanese decoration.

There were also great jewelers whose work, though not centrally of the Art Nouveau style, in some senses made it possible: Castellani, Giuliano and Falize, for example. Fortunato Pio Castellani (1793–1865) was fascinated by the past, and had been producing copies of ancient Greek, Roman and Etruscan jewelry in his Rome workshops since the 1820s. He achieved a revival of *plique-à-jour* enameling, which had been in decline since the sixteenth century but was to become the technique most favored by Art Nouveau enamelers. Fortunato retired in 1851, but his sons Alessandro and Augusto carried on the business, exhibiting some of their most remarkable work at London in 1862.

Castellani opened a workshop in 1860 at 13 Frith Street in London's Soho and sent Carlo Giuliano (1831–1912), one of their finest craftsmen, to manage it. Within a short time the atelier was taken over by Giuliano, who operated as a manufacturer, supplying the retail jewelry trade. By 1874 his artistic and financial success was such that he opened a magnificent retail shop at 115 Piccadilly. Giuliano's jewels were inspired by Renaissance principles, but within a refined idiom he was an adventurous and imaginative designer who incorporated

Colorplate 26.
The firm of J. Tostrup in Oslo excelled at the *plique-à-jour* technique. Torolf Prytz was in charge of artistic production, and he designed this magnificent goblet, which was made by Emil Saether, Tostrup's leading technician. Norwegian, 1900. Height: 32.5 cm. (12¾ in.). Kunstindustrimuseet, Oslo

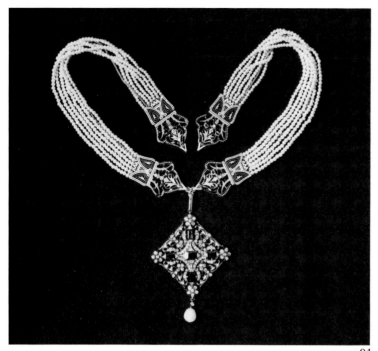

94

95

94.
Necklace and pendant brooch by Carlo Giuliano (active from 1861; d. 1912), a Neapolitan jeweler who achieved fame in London, where he opened a retail shop in Piccadilly. The necklace has six strands of seed pearls with gold attachments enameled in white. The pendant is set with peridots and diamonds, with a drop pearl. English, late nineteenth century. Cooper-Hewitt Museum, purchased in memory of Annie Schermerhorn Kane and Susan Dwight Bliss

95.
An oval gold locket of about 1869 decorated in the Japanese style with *cloisonné* enamel. Designed by Alexis Falize—it bears the initials "AF"—the work was executed by Antoine Tard for the firm of Falize. Objects such as this paved the way for the freedom of Art Nouveau. From the 1860s Oriental influences on works of art developed rapidly, asymmetry and fantasy in coloring becoming high fashion. The grasshopper on this locket has a blue body, mauve wings and lacquer-red legs, and is seen against a yellow sky. French, nineteenth century. Height: 6.1 cm. (2⅜ in.). Cleveland Museum of Art, purchase, Andrew R. and Martha Holden Jennings Fund

enamel into much of his work (plate 94). His sons Carlo and Arthur managed the firm from 1895 until it closed in 1914.

The French showed themselves especially sensitive to the prevailing influences. The Falize family, Alexis (1811–1898), his son Lucien and in turn his grandson André, made a specialty of enameling in the Oriental manner. Lucien was only twenty-three when he visited the 1862 London Exhibition. He returned to France captivated by the Japanese work he had seen. The family produced striking and delicate designs in both Japanese-style *cloisonné* enamels and Renaissance and neoclassical-type jewelry.

The Falize locket in plate 95 is fitted in a presentation case stamped inside the lid: "Tiffany & Co/550-Broadway-522/New York" (Tiffany moved to new premises in Union Square in 1870). At first the Falizes were manufacturing jewelers, supplying retail shops with their high-quality, superbly designed products. But the firm became known to a wider public as it began to show in its own right at important exhibitions such as the 1867 Paris Exposition de l'Union Centrale des Arts Décoratifs.

Eugène Fontenay (1823–1887), inspired by Castellani, produced a wonderful range of jewelry in neoclassical and Etruscan styles. Another Frenchman, Fernand Thesmar (1843–1912), a brilliant designer and enameler who worked for the Barbedienne factory, was also inspired by Japanese models (plate 96).

But the Paris Exposition of 1867 was most important in deriving fashionable aesthetic influences from the distant past. Work on the Suez Canal and the accompanying archeological excavations meant that in Paris at that time ancient Egypt was all the rage.

There were, then, many impetuses: from Castellani toward the ancient world; from China and Japan; from newly discovered Egypt; from Giuliano with his love of the Renaissance; and from the work of Falize and Thesmar.

In England, William Morris (1834–1896) initiated the Arts and Crafts movement, which in 1888 held its first exhibition at the New Gallery in London. C. R. Ashbee (1863–1942) founded the Guild and School of Handicrafts, and Henry Wilson (1864–1934), a leading designer, taught metalwork at the Royal College of Art. The famous store Liberty's, in London's Regent Street, was founded in 1875 by Sir Arthur Lazenby Liberty (1843–1917). This was the high point of the Aesthetic (or Art) movement, and the store became a vital focus for an unconventional generation of artists and craftsmen. All these influences combined to form a distinctive style, Art Nouveau. For a while enameling was transformed by exciting new ideas: a new art for the approaching new century, it must have seemed. Morris's principles went directly to nature as a source of ideas, shrugging off all aesthetic influences except some of the most remote and primitive: medieval and Celtic.

English art silver especially made use of enameling, on hand mirrors, candlesticks, presentation caskets, buckles and clasps, cigarette and cigar boxes, rose bowls, lamps and many other objects. Blues and greens were the predominant enamel colors; these were the easiest to fire and they also echoed the shades of the peacock, one of Art Nouveau's favorite symbols. As for personal jewelry, there was a far greater range than nowadays: tiepins, hatpins and hair combs are examples of almost-forgotten forms of adornment. In 1905 Charles

96.
A bowl with yellow flowers on a turquoise ground by Fernand Thesmar, exemplifying the influence Japanese works of art exerted on French artists in the nineteenth century. Bowls with allover *plique-à-jour* naturalistic flowers were a Japanese development. Although similar objects were made in Norway and Russia, their designs were more stylized and traditional. French, late nineteenth century. Diameter: 6.3 cm. (2½ in.). Metropolitan Museum of Art, New York, gift of Mrs. Charles Inman Barnard, 1905

Horner of Halifax opened a factory where enameled hatpins and pendants were made in Art Nouveau designs of startling asymmetry.

Many types of Art Nouveau objects were made by distinguished English craftsmen and craftswomen, and were briefly very popular. Kate Eadie and Phoebe Traquair produced painted enamels, as did the workshops of Omar Ramsden (1873–1939). Other important enamelers included Harold Stabler (1872–1945; plate 97), a versatile and influential craftsman; the Gaskin brothers of Birmingham; and, most accomplished of all, Alexander Fisher (see plate 93).

Fisher, an inspired teacher and brilliant artist, experimented boldly with techniques and colors, especially in the building up of many layers of different-colored translucent enamels. He also produced some dazzling *plique-à-jour* creations. But unfortunately enameling did not remain fashionable, nor did Art Nouveau seize and dominate the imagination of English designers for very long.

In France, however, it was a different matter. Jewelers there had always been bold and daring in their experiments, and during the 1860s they threw off the stranglehold of symmetry to revel in simplicity, spareness and lightness. The heavy, jewel-encrusted creations of mid-nineteenth-century taste were resoundingly rejected. The man who was to catch the feelings of the time and translate them into an aesthetic achievement was René Lalique (1860–1945; plate 98).

At sixteen Lalique was apprenticed to the jeweler Louis Aucoc in Paris, and later studied in England before executing his first designs as a freelancer. Then, in 1884, he took over the workshop of a jeweler who was retiring and immersed himself totally in the creation of a

97.
An Art Deco plaque in *cloisonné* enamel by Phoebe and Harold Stabler. Against a blue background a faun plays a pipe, surrounded by multicolored flowers. Artist, teacher, craftsman-potter, silversmith and enamelist, Harold Stabler was a central figure in the Arts and Crafts movement. He and his wife, Phoebe, produced a number of pottery figures and enamels. English, 1921. Mounted in a silver frame, width 7.5 cm. (2⅞ in.). Cooper-Hewitt Museum, gift of J. Lionberger Davis

new and distinctive art form. Many of Lalique's inspirations came from nature. He was particularly fond of autumnal images of fading leaves, browning bracken fronds, the first kiss of frost on the flower stalk: languor and luxuriant decay. All very suitable, obviously, for a sophisticated clientele that was growing tired of the endlessly repeated motifs of spring and summer. Lalique used *plique-à-jour* to create the illusion of the transparent membranes of nature—bat and butterfly wings, fish fins and leaves.

From 1895 Lalique introduced human figures into his designs, which caused a furor over whether this was simply bad taste or a daring and outrageous variation of what had been common practice for goldsmiths in the ancient world and during the Renaissance. Occasionally Lalique's imagination carried him further into the realms of the perverse, with fighting cocks spitting forth diamonds and monsters and exotic insects devouring one another. As with his interest in autumnal decay, Lalique shows here a distinctively *fin-de-siècle* fascination with the fringes of pain and cruelty. Yet the work is transformed, by the delicacy of line and the impassive faces, into designs of elegance and sophistication.

One of Lalique's most distinguished enamelers was Eugène Feuillâtre, who was also an accomplished sculptor and silversmith. He executed much of Lalique's finest *plique-à-jour* work until 1889, when he started on his own; Feuillâtre's enamels were featured at many major exhibitions.

A number of Lalique's contemporaries and disciples produced outstanding and original work, among them Ernest Vever and his sons

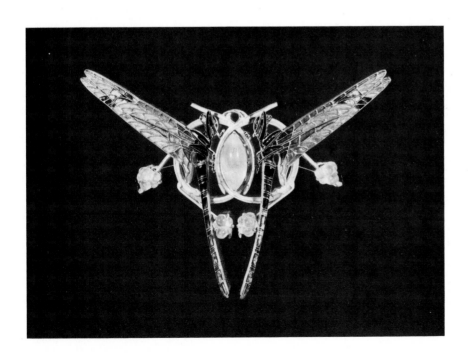

98.
A brooch in the form of a dragonfly by René Lalique in gold, *plique-à-jour* enamel, glass and carved moonstones. The translucent enamel wings shade from a pale gray-blue to an iridescent blue-green. French, nineteenth century. Height: 7.6 cm. (3 in.). Cleveland Museum of Art, purchase, Roberta Holden Bole Fund

Henri and Paul. Henri (1854–1942) wrote the standard text on French nineteenth-century jewelry. The Vevers employed the finest craftsmen, but their pieces, although among the best-designed Art Nouveau jewelry, lack the imaginative edge of wildness that renders Lalique's creations unique. Other notable enamelers of this period were Eugène Tourette, who was master enameler at the House of Vever, Paul Riquet (see colorplate 32), Alexandre Charlot de Courcy and Théophile Soyer.

In America the Tiffany family was alert to aesthetic developments in Europe. Louis Comfort Tiffany (1848–1933), who was an innovative artist in many fields, including glassmaking and jewelry design, was principally responsible for introducing Art Nouveau into the United States, and himself contributed toward it. In the late 1890s Tiffany worked with Julia Munson, producing bowls and vases in *repoussé* copper decorated with layers of opaque and translucent enamels, then given a final coat of iridescent enamel (plate 99). The distinctive Tiffany influence on design that blossomed under his guidance prospers to this day.

Enameling was used by other jewelers in the United States, among them Marcus & Company of New York, whose work incorporated vivid *plique-à-jour* enamel. In Boston, Elizabeth E. Copeland (colorplate 27)—a silversmith-enameler, as were so many of her European contemporaries—helped to establish the Arts and Crafts movement.

Colorplate 27.
A silver and enamel box by Elizabeth E. Copeland, whose work was influenced by primitive forms and natural motifs. Born in Boston, Miss Copeland studied in London before returning to her hometown, where she was associated with the Boston Society of Arts and Crafts Handicraft Shop. In 1916 she was awarded a medal for excellence by the Society. The box is enameled in blue, purple, yellow and green and is marked on the base in raised letters "E.C." American, Boston, 1907–16. Length: 16.5 cm. (6⅜ in.). Yale University Art Gallery, Mrs. Paul Moore Fund

99.
In the late 1890s, Louis C. Tiffany experimented with enameling at his home on 72nd Street, aided by two assistants; one of them was Julia Munson, the maker of this copper *repoussé* bowl. The design of plums, branches and leaves is enameled in iridescent tones of purple, brown, orange and green. Marked in incised script: "Louis C. Tiffany." American, New York, c. 1899–1910. Diameter: 24.2 cm. (9½ in.). Metropolitan Museum of Art, New York, gift of the Louis Comfort Tiffany Foundation, 1951

Colorplate 28.
A gilt-metal sweetmeat dish, in brilliantly
multicolored translucent *plique-à-jour*
enamel, depicting couples dancing in a
landscape with a windmill and, on the base,
vine leaves and grapes. Stamped on the base:
"Masriera Y Carreras." Luis de Masriera
(1872–1958) of Barcelona was an innovative
jeweler who studied in Geneva with Lossier,
a famous enameler. In 1915 the jewelry
firms of Carreras and Masriera (established
in 1750 and 1839, respectively) merged;
their enameling workshop is still active.
Spanish, 1915–30. Height: 12.1 cm. (4¾ in.).
Cooper-Hewitt Museum, gift of J. Lionberger Davis

Colorplate 29.
Mediterranée, a free-standing *plique-à-jour* enamel sculpture by Marit Guinness Aschan, an English artist-enameler who pioneered the development of this technique on the grand scale. Born in 1919, of Irish and Norwegian descent, Mrs. Aschan maintains a London studio in Chelsea. Height: 43.2 cm. (16⅞ in.). North Carolina Museum of Art, Raleigh, gift of Edward B. Benjamin, New Orleans and Greensboro, North Carolina

The influence of Art Nouveau spread to Belgium, where Philippe Wolfers (1858–1929) worked. A restless, exploratory designer, his flower motifs made use of ivory from the Congo. In Vienna, designs in *plique-à-jour* with a strong Art Nouveau tendency were developed; and in Spain, Luis de Masriera (1872–1958; colorplate 28) became known as an enameler whose work spanned the Art Nouveau and Art Deco periods. Norway's contribution to Art Nouveau enameling was outstanding: J. Tostrup of Oslo produced magnificent *plique-à-jour* bowls, lamps and vases (colorplate 26).

But within a few years of the Paris Exposition of 1900, Art Nouveau had had its day. In 1895 Lalique's perverse creations had shocked and delighted the senses; a decade later, it was all rather déjà vu. The style endured until World War I, but popular taste recoiled from the abandoned and writhing curves of Art Nouveau. For the next fifty years enameling, neglected and undervalued, would become the Cinderella of the arts.

Art Deco brought with it a hard-edged use of color in which the intricacies of enameling (as it had hitherto been employed) seemed out of place. Some artists ventured to interpret in enamel the cubist patterns and stylized designs of the period (plate 100). But the in-

100.
Cubism influenced enamel design during the Art Deco period. These vases from the renowned Fauré workshop typify the use of enamel on copper in the Modernist manner; translucent and transparent enamels graduate from a deep cherry red into a soft yellow and opalescent white. The taller vase is signed under the lip: "C. Fauré, Limoges, France." French, c. 1925. Height of taller vase: 23.7 cm. (9¼ in.). Cooper-Hewitt Museum

Colorplate 30.
Stefan Knapp (b. 1921) was the originator of enameled steel murals and for over twenty years has been a leading exponent of the art. His works are herculean: the enamel façade of Alexander's department store in Paramus, New Jersey, was the largest painting in the world when it was erected in 1960. It is five stories high and 98.4 m. (320 ft.) in length.

terior decorators of the 1920s, who set the style for everything that was *de rigueur* in fashionable homes, favored ceramics and lacquer. Jewelry design returned to discipline and symmetry and to the predominance of precious materials, influenced by Vienna and Berlin.

Colorplate 31.
Examples of the revival of the English eighteenth-century craft of enameling on copper: small boxes produced by the traditional method of transfer printing and overpainting in colors by hand.
Left: Detail from a Chinese nineteenth-century embroidery in the collection of the Metropolitan Museum of Art, New York.
Right and center: Replicas of Georgian enamels from the collections of two London museums—*An English Eighteenth-Century Garden with Classical Ruins*, from the British Museum, and *The Horse Guards*, from the Victoria and Albert Museum.
Bilston, South Staffordshire, 1981–2. Diameter of circular box: 5.7 cm. (2¼ in.)

Enamels Today   In the Far East the mass production of replicas of enamels from centuries past continues apace and, in some instances, the standard of work approaches that of the original antique models. Only in Japan, apparently, are artists experimenting with enameling for creative contemporary design.

In the West the potential of the medium is being fully exploited. Variations on techniques are being explored by jewelers and silversmiths to achieve imaginative, original effects.

At Limoges, ecclesiastical and secular enameled paintings are once more being produced, and the monks at Saint Martin de Ligugé are leaders of this movement. Following the sixteenth-century tradition, much of their subject matter has been the work of famous artists, among them Georges Rouault and Georges Braque. There are more than two hundred master enamelers listed as currently active in France, the majority of them in the Limoges area.

From Norway to Venezuela, throughout Europe and across America, creative enameling is flourishing, as countless exhibits in museums and art galleries testify (colorplate 29).

Vast enameled steel murals and sculptures are making a striking contribution to modern architecture. The leading exponent of this form of enameling is Polish-born Stefan Knapp (b. 1921), whose massive works have decorated more than 150 buildings in the United States alone (colorplate 30).

In England the traditional eighteenth-century art of making transfer-printed and painted small boxes and objects was revived in 1970 in Bilston in South Staffordshire, a leading center in Georgian times (colorplate 31). Apart from the use of electricity instead of solid fuel for heating the kilns, the manufacturing processes employed today are similar to those practiced two hundred years ago. Some of the enamel boxes currently being made in Bilston are decorated with designs after works by famous modern artists.

Contemporary enameling is, therefore, achieving success in several distinctly different directions: the proliferation of works by individual artist-enamelers throughout the Western world; the architectural application of the medium; and the revival of English enamel keepsakes and small collectors' pieces. Today, it seems, the future bodes well for enameling.

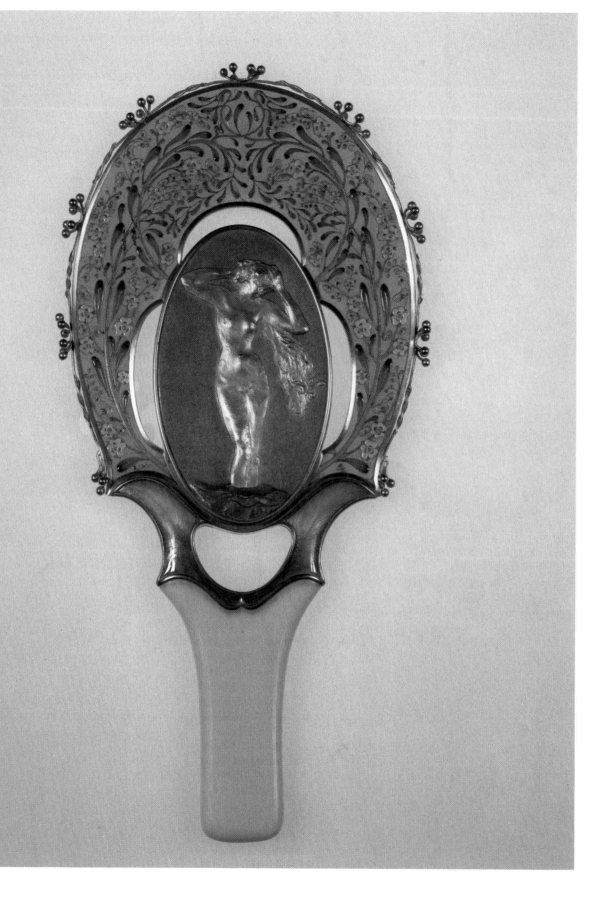

11 Advice for Collectors

It is beyond the reach of the average collector to aspire to rare specimens of museum caliber, and many of the enamels made before about 1800 are very expensive indeed. There are, however, countless enamels of many types that are both desirable and available at reasonable cost. The cardinal rule must be: collect what appeals to you. But a discriminating collection that explores the variations and contrasts to be found within a specific type or period of enamels is naturally of greater interest and value than a random number of totally unrelated objects.

Do not be deterred from buying a high-quality piece that has imperfections. Enameling is in any case an imprecise art, owing to the hazardous processes of manufacture, and blemishes are present on most examples; they are part of the character of objects made lovingly by hand centuries before the age of machine-made precision.

An antique enamel in pristine condition obviously is the most desirable. But is there a serious collector who—aside from its curiosity value—would prefer an unblemished nineteenth-century Viennese reproduction of a seventeenth-century ornament to a genuine eighteenth-century Battersea snuffbox, even with faults? Through years of use, the latter would show signs of wear, scratches on the underside and chips here and there on the corners, because its function was, after all, to be used.

There is also the controversial question of whether enamels should be restored. A. Kenneth Snowman, writing on the degree of restoration that is acceptable, says: "We are hardly surprised when we find that an old painting has sustained some sort of damage during its long life, and we have come to expect a picture restorer to make good the blemish as well as he is able. Nobody really wants hanging in his home a Renaissance portrait of a lady, however superbly painted, if her left

Colorplate 32.
A supreme example of the collaboration that existed between artists in Paris during the Art Nouveau period. This unique hand mirror was commissioned by patron of the arts and collector Baron Joseph Vitta, as a present for his mother. The work was supervised and probably designed by Félix Bracquemond (1833–1914), painter, etcher, ceramist and decorator. The figure of Venus Astarte, the Phoenician goddess of fertility and love, was by the sculptor Auguste Rodin (1840–1917); the plaquette was cast by the firm of Falize; and the translucent *plique-à-jour* enamel was by the master enamelist Paul Riquet. The mirror has an ivory handle and bears the monogram of Baron Vitta, the marks of Bracquemond and Riquet and the names of Rodin and Falize. French, dated 1900. Height: 32.2 cm. (12¾ in.). Cleveland Museum of Art, gift of Ralph King, by purchase

eye is missing. The same principle should naturally apply to the restoration of an old box. . . ."

Throughout history precious objects have been forged and copied, and enamels have not escaped. During the nineteenth century in particular, copying ancient and Renaissance artifacts became a considerable industry, and in most instances only an expert can recognize a fake. In the case of replicas of English eighteenth-century enamel boxes, made between about 1870 and the 1950s by the Parisian firm of Samson et Cie, identification of the fake is made easier by the existence of a small protruding metal lug just beneath the center of the outside hinge. This is never present on genuine English enamels.

But a collector need not seek only ancient or obscure examples. At present there is a worldwide renaissance in all aspects of enameling and a fascinating variety of work is being produced.

Collecting is most rewarding if the subject is studied in depth. Read the books; go to classes; scrutinize auction catalogues; talk to reputable dealers; visit museums and browse for hours. You will soon learn not only to treasure enamels but also to understand them.

Glossary

à fond vermiculé, ornamental engraving on a metal background composed of sinuous lines reminiscent of the tracks of worms.

basse-taille, the technique in which depressions in metal (normally gold or silver) are engraved, chased or carved into a pattern or design, then covered with translucent or transparent enamel.

blank, a piece of metal or an undecorated object, prepared or formed and ready for further processing.

champlevé, the technique in which a design is carved into metal and the hollows filled with enamel, the metal partitions dividing areas or colors of enamel.

chase, a method of working a pattern on metal whereby particles are not removed as they would be in engraving.

chinoiserie, European style of decoration using Chinese scenes or motifs.

cloisonné, the enameling technique in which strips of metal are attached at right angles to a metal base, forming tiny upstanding walls and thus creating cells (*cloisons*) to form a pattern; the cells are then filled with enamel and the surface can be polished until smooth.

(contre-émail) counter-enameling, enamel placed on the reverse side of metal that is enameled on the surface, to counteract the tension and contraction of the metal during firing.

cuprite glass, used to decorate Iron Age bronzes. Generally referred to as red enamel, the brilliant sealing-wax red is due to crystals of cuprous oxide in the composition.

émail en ronde bosse, the technique, also called encrusted enamel, in which enamel is applied to metal (usually gold) surfaces, either three-dimensional or in high relief.

en grisaille, decorative painting in various tones of gray to represent figures or objects in relief.

enlevage à l'aiguille, one of the earliest forms of painted enamel, whereby a layer of gold or white enamel is applied on very dark-colored enamel and then scraped away with a needle, enabling the dark background to show through to delineate a design.

en plein, the technique in which a field of enamel is floated onto gold or silver, creating an additional layer of its own rather than filling prepared recesses.

en résille sur verre, the technique in which a small piece of glass (normally blue or green) is engraved with a design, gold foil is placed into the recesses, and then enamel is placed on top of the foil. After firing, the foil remains visible around the edges of the enamel.

étui, a decorative container fitted with small personal accessories.

filigree enameling, the technique in which cells, or cloisons, are formed with twisted wire and enamel is inserted into the cells but does not fill them, as in *cloisonné* enameling. The twisted wires remain raised as part of the decoration.

flux, an almost colorless enamel, before oxides are added for coloring.

frit, enamel in the raw or ground state.

guilloché enameling, the technique in which enamel is applied to engine-turned surfaces.

millefiori, rods of different-colored glass bundled together, heated, and while malleable drawn out like chewing gum to a length of perhaps 9 to 12 meters (30 to 40 feet), thus reducing the cross-section. When cool, the slender glass rod is cut into thin slices, forming small multicolored flowered or mosaic platelets. For *millefiori* "enamel," a metal area is covered with the platelets and warmed just sufficiently for them to adhere to each other and the metal without being melted.

muffle, an enclosed compartment placed within a kiln or furnace to protect the piece being fired from debris and burning fuel as well as to ensure even distribution of heat.

paillons, (1) small slivers of thin gold or silver foil placed between layers of transparent or translucent enamel *or* (2) gold or silver cut-out and embossed encrustations applied to enamel or porcelain surfaces. (The word *paillons* also describes sequins or spangles used for the ornamentation of clothing.)

plique-à-jour (émail de plique), a backless *cloisonné* enameling technique that allows light to show through the colored transparent or translucent enamel within the cells (*cloisons*).

reliquary, a container for holy relics.

repoussé, metal raised into ornamental relief by means of hammering from the reverse side.

reserve, an area left plain within a decorated background, ready to receive a different form of decoration.

skan enameling, a technique that combines the *cloisonné* and filigree methods, using formal floral patterns enclosed by "pearl" borders composed of encrusted drops of white enamel.

slurry, an opaque liquid mixture of enamel powder, water, and less than 5 per cent clay with a minimal amount of a chemical added to render the liquid viscous.

spike oil, a distillation made from *Lavandula spica* (lavender), employed in enamel painting, which enables the enamel to spread but which burns away during firing.

transfer printing, a process by which designs are printed onto ceramic or vitreous surfaces.

Reading and Reference

General

AMIRANSHVILI, SHALVA. *Georgian Metalwork.* London: Paul Hamlyn, 1971.

BARSALI, ISA BELLI. *European Enamels.* London: Paul Hamlyn, 1969.

BATES, KENNETH F. *Enamelling: Principles and Practice.* New York: Funk & Wagnalls, 1974.

BECKER, VIVIENNE. *Antique and 20th Century Jewellery.* London: N.A.G. Press, 1980.

BENJAMIN, SUSAN. *English Enamel Boxes.* New York: Viking Press, 1978.

BIRDWOOD, GEORGE C. M. *The Industrial Arts of India.* London: Chapman & Hall, 1880.

BRAMSEN, BO. *Nordiske Snüsdaser.* Copenhagen: Politikens Forlag, 1965.

BRONSTEIN, LEO. *A Survey of Persian Art.* London: Oxford University Press, 1964–5.

CELLINI, BENVENUTO. *Treatises of Goldsmithing and Sculpture,* trans. by C. R. Ashbee. New York: Dover Publications, 1966.

CHAMOT, M. *English Mediaeval Enamels.* London: Ernest Benn, 1930.

COLDING, TARBEN HOLCK. *Danish Miniaturists.* Copenhagen: Gyldenalske Boghandel Nordisk Forlag, 1948.

FISHER, ALEXANDER. *The Art of Enamelling.* London: Bradbury, Agnew, 1906.

GARNER, SIR HARRY. *Chinese and Japanese Cloisonné Enamels.* London: Faber & Faber, 1976.

GAUTHIER, MARIE MADELEINE, AND GENEVIEVE FRANÇOIS. *Mediaeval Enamels, Masterpieces from the Keir Collection.* London: British Museum Publications Ltd., 1981.

GLUCK, JAY, AND SUMI HIRAMOTO. *A Survey of Persian Handicrafts.* Tehran: Bank Melli, 1977.

HAMILTON, G. H. *History of Art and Architecture of Russia.* Harmondsworth, Middlesex: Pelican History of Art, 1975.

HILDBURGH, W. L. *Mediaeval Spanish Enamels.* London: Oxford University Press, 1936.

HINKS, PETER. *Nineteenth Century Jewellery.* London: Faber & Faber, 1975.

JACOB, S. S., AND T. H. HENDLEY. *Jeypore Enamels.* London: W. Griggs, 1886.

JAMES A. DE ROTHSCHILD COLLECTION AT WADDESDON MANOR: *Limoges Enamels,* Madeleine Marcheix; *Painted Enamels Other Than Limoges,* R. J. Charleston. The National Trust, London, 1977.

Jewellery Through 7000 Years. London: British Museum Publications Ltd., 1978.

LABARTE, JULES. *Recherches sur la peinture en émail.* Paris, 1856.

LATIF, MOMIN. *Mughal Jewels.* Brussels: Société Générale de Banque, 1982.

MARYON, HERBERT. *Metalwork and Enamelling.* New York: Dover Publications, 1971.

MIHALIK, SANDOR. *Old Hungarian Enamels.* Budapest: Corvina Press, 1961.

POLAK, ADA. *Norwegian Silver.* Oslo: Dreyers Forlag, 1972.

RICKETTS, HOWARD. *Objects of Vertu.* London: Barrie & Jenkins, 1971.

ROSS, MARVIN C. *The Art of Karl Fabergé and His Contemporaries.* Norman: University of Oklahoma Press, 1965.

SNOWMAN, A. KENNETH. *Eighteenth Century Gold Boxes of Europe.* Boston: Boston Book & Art Shop, 1966.
——. *The Art of Carl Fabergé.* London: Faber & Faber, 1968.

THEOPHILUS. *On Divers Arts,* trans. John A. Hawthorne and Cyril Stanley Smith. New York: Dover Publications, 1963.

VERDIER, PHILIPPE. *The Frick Collection,* Vol. VIII, *Limoges Painted Enamels.* New York: The Frick Collection, 1977.
——. *Painted Enamels of the Renaissance, Walters Art Gallery.* Baltimore: published by the Trustees, 1967.

VEVER, HENRI. *Bijouterie Française,* Vol. III. Florence: Studio per edizioni scelte, 1906.

WESSEL, KLAUS. *Byzantine Enamels.* Shannon: Irish University Press, 1969.

Guides, Articles and Papers

BEARD, CHARLES R. "Surrey Enamels of the Seventeenth Century." *The Connoisseur,* October 1931.

BENJAMIN, SUSAN. "Designs on Some Eighteenth Century English Enamels." *The Connoisseur,* January 1979.

BOINET, AMEDÉE, "Exhibition of Limousin Enamels at the Museum of Limoges." *The Connoisseur,* December 1948.

DODDS, LADY. "Persian Enamels." *Apollo,* September 1960.

FORSYTH, WILLIAM H. "Around Godefroid de Claire." *The Metropolitan Museum of Art Bulletin,* Vol. 24 (June 1966).
——. "Mediaeval Enamels in a New Installation." *The Metropolitan Museum of Art Bulletin,* Vol. 4, no. 9 (May 1946).
——. "Provincial Roman Enamels Recently Acquired." *The Metropolitan Museum of Art Bulletin,* Vol. 2, no. 32 (1950).

Guide to Anglo-Saxon Antiquities. London: British Museum, 1923.

Guide to Early Christian and Byzantine Antiquities. London: British Museum, 1903.

Guide to Early Iron Age Antiquities. London: British Museum, 1925.

HARADA, JIRO. "Japanese Art and Artists of Today." *Studio,* Vol. 53 (September 1911).

HAWLEY, HENRY H. "A Falize Locket." *The Cleveland Museum of Art Bulletin,* Vol. 66 (September 1979).

HAYWARD, J. F. "Salomon Weininger, Master Faker." *The Connoisseur,* November 1974.

HUGHES, M. J. "A Technical Study of Opaque Red Glass of the Iron Age in Britain." *Proceedings of the Prehistoric Society,* Cambridge, England, 1972.

JENYNS, SOAME. "The Problem of Chinese Cloisonné Enamels." *The Oriental Ceramic Society,* 1950.

MILLIKEN, WILLIAM M. "A Danish Champlevé Enamel." *The Cleveland Museum of Art Bulletin,* Vol. 36 (June 1949).

MUNN, GEOFFREY. "The Giuliano Family." *The Connoisseur,* November 1975.

OMAN, C. C. "A Note on Designs by Valentin Sezenius." *Apollo,* Vol. VI (1927).

WEISBERG, GABRIEL P. "Baron Vitta and the Bracquemond/Rodin Mirror." *The Cleveland Museum of Art Bulletin,* Vol. 66 (November 1979).

Some Public Collections of Enamels

UNITED STATES

Baltimore:	The Walters Art Gallery
Boston:	Museum of Fine Arts
Chicago:	Art Institute of Chicago
Cincinnati:	The Taft Museum
Cleveland:	Cleveland Museum of Art
New York City:	The Brooklyn Museum
	Cooper-Hewitt Museum, the Smithsonian Institution's National Museum of Design
	The Frick Collection The Metropolitan Museum of Art
Richmond:	Virginia Museum of Fine Arts
Washington, D.C.:	Dumbarton Oaks Collection Marjorie Merriweather Post Collection, Hillwood
	Smithsonian Institution
	National Gallery of Fine Arts
Williamsburg, Va.:	Governor's Palace, Colonial Williamsburg

OTHER

Amsterdam:	Rijksmuseum
Barcelona:	Museo d'Arte de Cataluña
Bedford, England:	Cecil Higgins Art Gallery
Birmingham, England:	The City Museum and Art Gallery
Budapest:	Hungarian National Museum
Cambridge:	Fitzwilliam Museum
Copenhagen:	The Royal Collection at Rosenbörg
Darmstadt:	Grossherzogliche Porzellansammlung
Dublin:	National Museum of Ireland
Edinburgh:	Royal Scottish Museum National Museum of Antiquities
Florence:	Museo Nazionale Bargello Museo Stibbert
Leningrad:	State Hermitage Museum The Russian Museum
London:	British Museum London Museum Victoria and Albert Museum
	Wallace Collection
Luton, England:	Wernher Collection, Luton Hoo
Madrid:	Museo Lazaro Galdiano
Manchester:	City Art Gallery
Milan:	Museo Poldi Pezzoli
Moscow:	The Armory Museum of the Kremlin
Munich:	Schatzkammer der Residenz, Munich
Naples:	Museo Nazionale
Oslo:	Kunstindustrimuseet
Oxford:	Ashmolean Museum
Paris:	Musée du Louvre Musée du Cluny
	Musée des Arts Décoratifs Musée Cognac-Jay
Rome:	Vatican Museum
Stockholm:	Nationalmuseum
Stuttgart:	Württembergische Landesmuseum
Vienna:	Kunsthistorisches Museum
Wolverhampton:	Bantock House Museum

Index

Acknowledgments

Cooper-Hewitt staff members have been responsible for the following contributions to the series: concept, Lisa Taylor; administration, Christian Rohlfing; coordination, David McFadden and Nancy Akre. In addition, valuable help has been provided by S. Dillon Ripley, Joseph Bonsignore, Susan Hamilton and Robert W. Mason of the Smithsonian Institution, as well as by Gloria Norris, Edward E. Fitzgerald, Madeleine Karter, Neal Jones and the late Warren Lynch of Book-of-the-Month Club, Inc.

Credits

The Ashmolean Museum, Oxford: plate 19. The Avery Brundage Collection, San Francisco: color 10. Bantock House Museum, Wolverhampton, England: color 25; plates 2, 78, 83, 84, 86, 88, 91. Susan Benjamin (Prudence Cuming Associates Limited, photographer), London: color 8, 9, 31. The Trustees of the British Museum, London: color 3, 5, 13, 18, 24; plates 9, 11, 13, 61, 66, 76, 77. Brooklyn Museum: plates 26, 27, 29, 32. Cleveland Museum of Art: color 32; plates 23, 24, 50, 51, 64, 95, 98. Cologne Cathedral (Das Rheinische Bildarchiv): plate 53. Cooper-Hewitt Museum (Scott Hyde, photographer), New York: color 28; plates 30, 35, 41, 87, 94, 97, 100. Fitzwilliam Museum, Cambridge: plate 45. The FORBES Magazine Collection, New York (Larry Stein, photographer): color 14. Copyright of the Frick Collection, New York: color 2. Gloucester City Museum and Art Gallery, Gloucester,

England: plate 12. Gyor Cathedral (Corvina Archives, Budapest, Hungary): color 20. *Harper's* magazine (Scott Hyde, photographer), New York: endpapers. Hungarian National Museum, Budapest: plate 73. Stefan Knapp, London: color 30. Kunstgewerbe Museum, Berlin: plate 47. Kunstindustrimuseet, Oslo: color 26; plate 48. Metropolitan Museum of Art, New York: color 1, 23; plates 14, 15, 17 and 18, 33, 52, 54, 58, 89, 96, 99. Minneapolis Institute of Arts: plate 34. Musée de Antiquités Nationales, Saint-Germain-en-Laye, France: plate 10. Musée de Cluny, Paris: plates 55, 56. Musée Pasteur, Paris (copyright Kunstindustrimuseet, Oslo): plate 49. Museum of Fine Arts, Boston: plate 25. National Museum of Antiquities of Scotland, Edinburgh: plate 75. National Museum of Ireland, Dublin: plate 74. Collection of the National Palace Museum, Taipei, Taiwan, Republic of China: plate 31. Nationalmuseum, Stockholm: plate 44. North Carolina Museum of Art, Raleigh: color 29. Residenz der Schatzkammer, Munich: color 16. Rijksmuseum, Amsterdam: plates 68, 69. Rosenbörg Castle, Copenhagen: plate 46. The Royal Treasury, Stockholm: color 15, plate 43. Soyer Factory (Metiers d'Art, photographie Stanislas de Grailly), Moulin de Saint Paul, Condat-sur-Vienne: plate 1. St. Mark's Cathedral, Venice (Vision International, London): color 4. Tiroler Landesmuseum Ferdinandeum, Innsbruck: plate 20. Toledo Museum of Art, Ohio: plate 65. University of Pennsylvania Museum, Philadelphia: plate 28. Virginia Museum of Art, Richmond: plate 42. Victoria and Albert Museum, Crown Copyright, London: frontispiece, color 6, 7, 11, 12; plates 8, 22, 57, 67, 70, 71, 72, 79 and 80, 81 and 82, 85, 92, 93. Waddesdon Manor, Aylesbury, Buckinghamshire: color 19; plate 90. Wallace Collection, London: color 21, 22; plates 62 and 63. Walters Art Gallery, Baltimore: color 17; plates 7, 16, 36, 37, 38, 39, 40, 59, 60. Wawel State Collection of Art, Cracow: plate 21. Yale University Art Gallery, New Haven: color 27.